Two Variable Algebra and Quadratics

David Eastwood

PARTRIDGE

Copyright © 2019 by Brain Based Education.
Math Without Calculators
Algebra Two
Medina, Ohio USA

ISBN: Softcover   978-1-5437-0556-0

All rights reserved. No part of this book may be used or reproduced by any means, graphic, electronic, or mechanical, including photocopying, recording, taping or by any information storage retrieval system without the written permission of the author except in the case of brief quotations embodied in critical articles and reviews.

Summary: "This book Is for students who want to learn math, especially those who want to learn it without calculators. It is filled with suggestions and problems to review it."

Because of the dynamic nature of the Internet, any web addresses or links contained in this book may have changed since publication and may no longer be valid. The views expressed in this work are solely those of the author and do not necessarily reflect the views of the publisher, and the publisher hereby disclaims any responsibility for them.

Print information available on the last page.

To order additional copies of this book, contact
Partridge India
000 800 10062 62
orders.india@partridgepublishing.com

www.partridgepublishing.com/india

6. Two Variable Algebra

1. **Graph Points and Lines**...............1
   - 1-1 Graph Points   1
   - 1-2 Find Points, Draw Lines   3
   - 1-3 Y intercept   5
   - 1-4 Line Predicting   7
   - 1-5 Review Problems   9

2. **Slope/Slope Intercept Form**.......11
   - 2-1 Slope   11
   - 2-2 Slope Intercept Formula   13
   - 2-3 Backwards   15
   - 2-4 Slope Formula   17
   - 2-5 Quad Formula   19
   - 2-6 Review Problems   21

3. **Move and Measure Lines**..........23
   - 3-1 Midpoint/ Line on Graph   23
   - 3-2 Distance Formula   25
   - 3-3 2 out of 3 Rule/ Parallel   27
   - 3-4 Perpendicular Line   29
   - 3-5 Review Problems   31

4. **Two Variable Story Problems**...33
   - 4-1 Change 1 to 2 Variables   33
   - 4-2 How Y intercept Changes It   35
   - 4-3 Percent Equations   37
   - 4-4 Add Rate Formulas   39
   - 4-5 Review Problems   41

5. **Using Two Variable Equations**..43
   - 5-1 X Intercepts Story Problems   43
   - 5-2 Both Intercepts Story Prob   45
   - 5-3 Profitability Problems   47
   - 5-4 How to Say and Make Prob   49
   - 5-5 Review Problems   51

6. **Distance and Rate Problems**.....53
   - 6-1 Change Feet/sec to Mph   53
   - 6-2 How Distance Makes Rate   55
   - 6-3 Percent of a Rate Formula   57
   - 6-4 Review Problems   59

7. **Rate and Acceleration Prob**......61
   - 7-1 What is Acceleration?   61
   - 7-2 Acceleration Equation   63
   - 7-3 Average Velocity/ Jet Prob   65
   - 7-4 Review Problems   67

8. **One/Two Variables Inequalities**.69
   - 8-1 One Variable Problems   69
   - 8-2 Two Variable Problems   71
   - 8-3 Story Problems   73
   - 8-4 Review Problems   75
   - 8-5 And/ Or Problems   77
   - 8-6 2 Step Problems   79
   - 8-7 Review Problems Pt 2   81

9. **One/Two Step Absolute Var**.......83
   - 9-1 Absolute Value Expressions   83
   - 9-2 Absolute Value Equations   85
   - 9-3 Two Variable Inequalities   87
   - 9-4 Review Problems   89
   - 9-5 Absolute Value   91
   - 9-6 Y intercepts   93
   - 9-7 Story Problems   95
   - 9-9 Review Problems   97

10. **Function Basics**........................99
    - 10-1 What is a Function?   99
    - 10-2 Domain Range   101
    - 10-3 Four Operations   103
    - 10-4 Transformations   105
    - 10-5 Review Problems   107

11. Piecewise Functions 109

11-1 Piecewise Functions	109
11-2 Using And Inequalities	111
11-3 Discontinuities	113
11-4 Story Problems	115
11-5 Review Problems	117

12. Simultaneous Equations 119

12-1 Simultations	119
12-2 Eliminations	121
12-3 Story Problems	123
12-4 Story Problems Pt 2	125
12-5 Review Problems	127

7. Quadratic Algebra

1. **Basic Quadratics and Graphs...129**
   - 1-1 Begin Quadratics   129
   - 1-2 How Quadratics Graph   131
   - 1-3 How A terms Graph   133
   - 1-4 Positive Quadratic Equations   135
   - 1-5 Review Problems   137

2. **How Trinomials Make Graphs...139**
   - 2-1 Begin Negatives Signs   139
   - 2-2 Axis of Symmetry   141
   - 2-3 How Trinomials Work   143
   - 2-4 Review Problems   145

3. **Multiplying Binomials...147**
   - 3-1 Name Different of Equations   147
   - 3-2 Multiply Negative Signs   149
   - 3-3 How A tems Multiply   151
   - 3-4 A 2nd Way To Multiply   153
   - 3-5 Review Problems   155

4. **Factoring Trinomials...157**
   - 4-1 Common Equations   157
   - 4-2 Factor 1 and 2 Equations   159
   - 4-3 Factor with A terms   161
   - 4-4 Review Problems   163

5. **Factor/ Finding X intercepts...165**
   - 5-1 Factror Negative Equations   165
   - 5-2 Negative Equations   167
   - 5-3 First Term Negatives   169
   - 5-4 X intercepts   171
   - 5-5 Easy Graphs   173
   - 5-6 Review Problems   175

6. **Complete the Square...177**
   - 6-1 Complete the Square   177
   - 6-2 Shorten Complete the Square   179
   - 6-3 Review Problems   181

7. **Quadratic Formula...183**
   - 7-1 Discrimanent Positives   183
   - 7-2 Quadratic Formula   185
   - 7-3 Negatives   187
   - 7-4 Review Problems   189

8. **Consecutive Num/Revenue Pr.191**
   - 8-1 Consecutive Numbers   191
   - 8-2 Revenue Problems   193
   - 8-3 Story Problems   195
   - 8-4 Inequality Problems   197
   - 8-5 Review Problems   199

9. **Falling Objects...201**
   - 9-1 Build an Arch Problem   201
   - 9-2 Real Graphs with B/C Terms   203
   - 9-3 Use Axis of Symmetry   205
   - 9-4 Review Problems   207

10. **Quick Equations...209**
    - 10-1 Quick Formula   209
    - 10-2 Complete the Square   211
    - 10-3 Shorten Complete the Square   213
    - 10-4 Review Problems   215

11. Quadratic Distance Formula......217

- 11-1 Quadratic Distance Formula   217
- 11-2 Time for Slowing Down   219
- 11-3 Connect Gravity/Planets   221
- 11-4 Rate from Acceleration   223
- 11-5 Review Problems   225

12. Work and Stream Problems........227

- 12-1 Same Denominators   227
- 12-2 Work Problems   229
- 12-3 Stream Problems   231
- 12-4 Centripetal Acceleration   233
- 12-5 Review Problems   235

13. Compound Interest......................237

- 13-1 All the Money Formula   237
- 13-2 Compounded Interest   239
- 13-3 Review Problems   241
- 13-4 Ready Made Equations   243
- 13-5 Review Problems   245

14. Pythagorean Th and Circles.......247

- 14-1 Pythagoreans Formula   247
- 14-2 Story Problems   249
- 14-3 Pythagorean and Circle   251
- 14-4 Simultaneous Equations   253
- 14-5 Review Problems   255

15. Science/Geometry Problems......257

- 15-1 Tank Problem   257
- 15-2 Garden and Pan Problem   259
- 15-3 Review Problems   261
- 15-4 Resistance Problem   263
- 15-5 Parachute/ Pendulum Pr   265
- 15-6 Review Problems   267

Ch 4 Ls 1: 2 Steps to Carry Addition   33

Why we're different!!!

_____ Front ____ / 8   Back ____ / 27   Rev ____ / 20   T / 53 _____
　　　　Name　　　　　　　　　　　　　　　　　　　　　　　　　　　　Checker

#1 1. When do you carry in math? _____

2. What does 6 + 6 carry? $\begin{array}{r}26\\+\ 6\end{array}$ _____

3. What's the 2nd step to carry? _____

4. Why don't you have to carry twice with 100s? _____

> 3. The checker signs here when they're done.

> 1. Student makes sure these are filled out in class.

#2 1. You know 3 + 9 is 12. What's the 1st step to carry?

$$\begin{array}{r}13\\+\ 9\\\hline 2\end{array}$$

Carry a ____.

$$\begin{array}{r}{}^1 13\\+\ 9\\\hline 2\end{array}$$

　　tens ones

> 2. Checker quizzes the student on the front page.

2. What's the teen fact?

$$\begin{array}{r}15\\+\ 9\end{array}$$

5 + 9 = ____

Carry a 10. How many 10s in all?

3. Find the teen fact.

$$\begin{array}{r}34\\+\ 8\end{array}$$

4 + 8 = ____

Carry a 10. How many 10s in all?

Student is honest about whether they used a calculator or not. → Calculator? yes no

Review 1. When do you carry in math? _____ Calculator?
        2. Name 2 steps to carry. _____ yes no
        3. Why don't you have to carry twice with 100s? _____

**Student is quizzed on these Qs.
(Teacher option)**

It works!!!

PLUS These explore these pages...

1. **What's Happening in Algebra** The student decides what's happening in algebra.

2. **Simplified Algebra** means it's easier to make equations, not just solve equations someone else has made.

3. **Plenty of story problems!!!**

It's written as Math Without Calculators, but you need to make it happen.

**You can do this!
Mr David Eastwood**

Rules Sheet

As you go through the lessons you will see some rules that you are unfamiliar with, so I developed this paper to help you understand them.

1. Ma Rule: The intials are from the rule itself, "Multiply same bases, Add the exponents." You probably know the Product of Powers Property.
   Example: $2^2 \times 2^4 = 2^6$

2. Me Rule: These initials are also from the rule, "Multiply the numbers, Exponents stay the same." (E shows that exponents stay the same.)
   Example: $3^4 \times 5^4 = 15^4$

3. DS Rule: Again, the initials show the rule. "Divide same bases, Subtract the Exponents."
   Example: 4^5 divided by $4^2 = 4^3$

4. Simpliify: No need for initials here, just simplify.
   Example: 20^3 divided by $4^3 = 5^3$

5. Subtraction: 2. Count Up Subtraction: Example: 23 - 9   Count Up 9 to 10 and subtract, it's 13. Add the 1 from the bottom. It's 14.

These are the name changes up until now. Any others?

Ch 1 Ls 1 Find points using 2 variables. 1

_____ #1 #2 ____/ 11 #3 ____/ 2 R ____/ 9 T ____/ 22 _____
           Name   Checker

#1 1. What do the variables x and y stand for? _____

2. In the point (3, 2), what do 3 and 2 stand for? _____

3. Find each point.

Point (___, ___)   Point (___, ___)

#2 1. How are the 4 quadrants labeled? _____

2. Which quadrants are negative X's? _____

3. Which quadrants are negative Y's? _____

4. What are the points and quads?

A (___, ___) Quad ___   B(___, ___) Quad ___

5. Find the points and quads.

A (___, ___) Quad ___   B(___, ___) Quad ___

#3 Find the points. How is the 2nd point different from the 1st?   Calculator?
      yes no

1.   A is (__, __)

       B is (__, __)

   A is _____

   _____

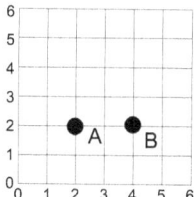

2.   A is (__, __)

       B is (__, __)

   A is _____

   _____

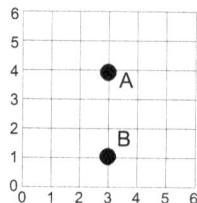

Review 1. What do the variables x and y stand for? _____   Calculator?
2. In the point (3, 2), what do 3 and 2 stand for? _____   yes no
3. How are the 4 quadrants labeled? _____
4. Which quadrants are negative X's? _____
5. Which quadrants are negative Y's? _____

What are the points and quads?

6.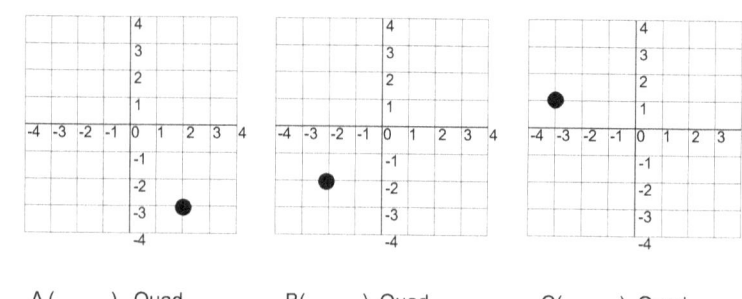

A (__, __) Quad ___   B(__, __) Quad ___   C(__, __) Quad ___

Each lesson has a quiz.

Ch 1 Ls 2 Find points from equations. 3

_____ #1 #2 ____ /10 #3 ____ / 4 R ____ / 9 T ____ / 23 _____
Name   Checker

#1 1. $y = x + 1$ How do you make a point from an equation? _____
    2. What's the name for the variable you put numbers in for? _____
    3. What's the name for the variable that shows the answers? _____
    4. How do you write it when you put a negative in for a variable? _____

    5. Put 5 in for X. Solve for Y. $x + 4 = y$

                                        ___ $+ 4 = y$ Find what Y is.
                                                           What point is it?

                                        ___ $= y$ (___, ___)

    6. Put 2 in for X. Solve for Y. $3x + 4 = y$

                                        ___ $+ 4 = y$ Find what Y is.
                                                           What point is it?

                                        ___ $= y$ (___, ___)

    7. Put 3 in for X. Solve for Y. $\frac{1}{2} x - 1 = y$

                                        ___ $- 1 = y$ Find what Y is.
                                                           What point is it?

                                        ___ $= y$ (___, ___)

#2 1. How do you graph a line? _____
    2. How many answers are on a line? _____
    3. Find X is 0 and 3. Is it positive or negative? $x + 5 = y$

                    x is 0 ___ , ___
                    x is 3 ___ , ___

                         Circle
                    Positive   Negative

4.

#3 Find each point from equations.   Calculator?
                                              yes   no

1.   x + 4 = y for x is 3   x - 1 = y for x is 5
     ___ + 4 = y   ___ - 1 = y
     ___ = y (___, ___)   ___ = y (___, ___)

2.   x - 3 = y for x is - 5   x + 2 = y for x is 7
     ___ - 3 = y   ___ + 2 = y
     ___ = y (___, ___)   ___ = y (___, ___)

Review 1. y = x + 1 What's the 1st step to get a point from an equation? _____   Calculator?
                                                                                            yes   no
2. After you put a number in for X, how does it make a point? _____
3. How do you write it when you put a negative in for a variable? _____
4. How do you graph a line? _____
5. How many answers are on a line? _____

Find 2 points and graph these equations.

6.

x + 2 = y

x is 0, ___

x is 3, ___

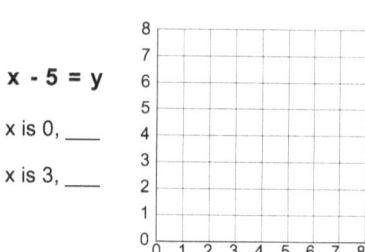

x - 5 = y

x is 0, ___

x is 3, ___

7.

x - 3 = y

x is 0, ___

x is 3, ___

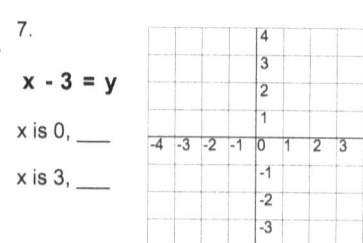

x + 1 = y

x is 0, ___

x is 3, ___

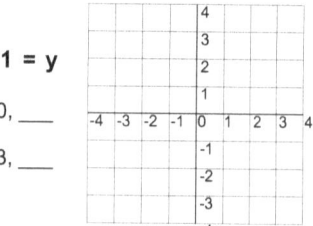

Ch 1 Ls 3 Use a y intercept to make a graph. 5

_____ #1 #2 ____ / 9 #3 ____ / 4 R ____ / 7 T ____ / 20 _____
Name   Checker

#1 1. What is a Y intercept? _____
    2. What is the Y intercept in $y = x + 1$? _____
    3. What makes a negative slope? _____
    4. How does just a Y or X (like x = 2) make a graph? _____

#2 1. How do you solve for the Y intercept? $y = 2x + 1$

                    Put ___ in for X. $y = 2$___$+ 1$ Finish it.

                                                                    What's the
                                                 _____   y intercept?

                    Y intercept is ___   _____

    2. How do you solve for the Y intercept? $y = -3x - 4$

                    Put ___ in for X. $y = -3$___$- 4$ Finish it.

                                                                    What's the
                                                 _____   y intercept?

                    Y intercept is ___   _____

    3. Solve for X is 8. What point is it? $y = -x - 3$ for x is 8

                                                 $y = $___$- 3$
                                                 $y = $___   Point is ___, ___.

    4. Solve for X is 6. Find the point. $y = -x - 1$ for x is 6

                                                 $y = $___$- 1$
                                                 $y = $___   Point is ___, ___.

    5. Solve for X is - 5. Find the point. $y = -x - 2$ for x is - 5

                                                 $y = $___$- 2$
                                                 $y = $___   Point is ___, ___.

#3 Decide the Y intercept and if slope is negative or positive.   Calculator? yes no

1.
Y intercept
0, _____

Positive
Negative

Y intercept
0, _____

Positive
Negative

2.
Y intercept
0, _____

Positive
Negative

Y intercept
0, _____

Positive
Negative

Review
1. What is a Y intercept? _____   Calculator? yes no
2. What is the Y intercept in $y = x + 1$? _____
3. What makes a negative slope? _____
4. How does just a Y or X (like $x = 2$) make a graph? _____

Graph these equations.

5. $y = -x + 2$ $y = x - 3$ $y = x + 1$

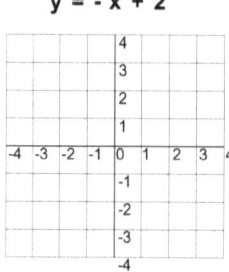
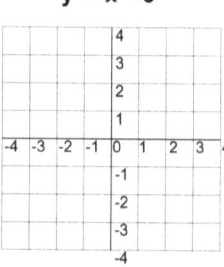
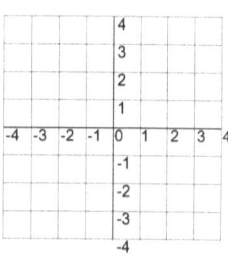

Ch 1 Ls 4 Predict graphs and equations. 7

_____ #1 #2 ____/ 5   #3 ____/ 4 R ____/ 5 T____/ 14 _____
                Name   Checker

#1 1. Name 2 parts of an equation that predict a line. _____
       2. What part of the equation starts the line? _____

#2 1. Make an equation. Find
        the slope and Y intercept.

Negative   Positive   Slope ____   Y int ____   y = _____

2. Make an equation with
   y intercept and slope.

Negative   Positive   Slope ____   Y int ____   y = _____

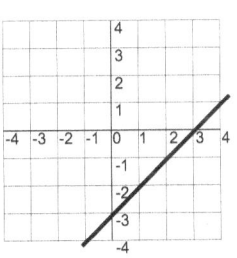

3. Make an equation.

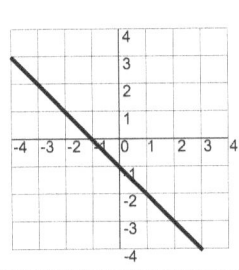

Negative   Positive   Slope ____   Y int ____   y = _____

#3 Make an equation.   Calculator? yes no

1.

y = _____

Y intercept

_____

Positive
Negative

y = _____

Y intercept

_____

Positive
Negative

2.

y = _____

Y intercept

_____

Positive
Negative

y = _____

Y intercept

_____

Positive
Negative

Review 1. Name 2 parts of an equation that predict a line. _____   Calculator? yes no

2. What part of the equation starts the line? _____

Fill in the graph for each equation.

3.   y = x - 1   y = - x + 2   y = - x - 3

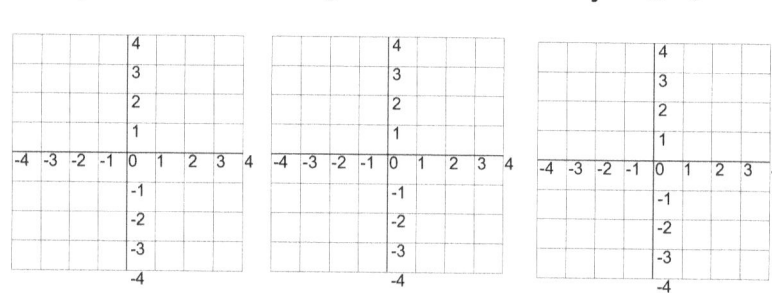

Problem Review 9

_____ #1 #2 ____/ 13 #3 #4 ____/ 8 T ____/ 21
                Name

#1 1. Number Grid _____
    2. X Axis _____
    3. Y Axis _____
    4. 2 Variable Algebra _____
    5. Y intercept _____
    6. Negative Slope _____

#2 What are the points and quads?   Calculator?
                                                       yes   no

1.
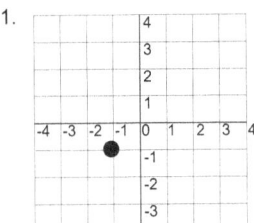
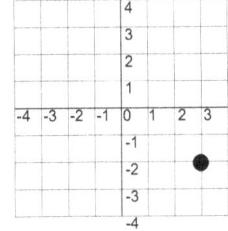
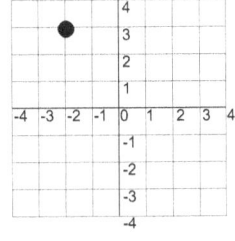

A(__, __) Quad ____   B(__, __) Quad ____   C(__, __) Quad ____

Draw the line in.

2.
x + 1 = y

0, ____
3, ____
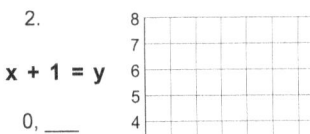

x + 3 = y

1, ____
3, ____
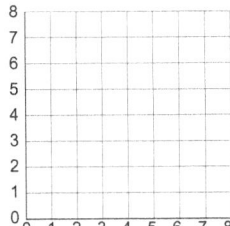

3.
x + 2 = y

- 1, ____
2, ____
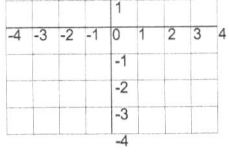

x - 3 = y

- 2, ____
3, ____
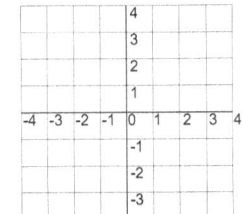

#3 Decide the Y intercept and if slope is negative or positive. Calculator? yes no

1.
Y intercept
0, _____

Positive
Negative

Y intercept
0, _____

Positive
Negative

2.
Y intercept
0, _____

Positive
Negative

Y intercept
0, _____

Positive
Negative

#4 Make an equation. Calculator? yes no

1.
y = _____

Y intercept
_____
Positive
Negative

y = _____

Y intercept
_____
Positive
Negative

2.
y = _____

Y intercept
_____
Positive
Negative

y = _____

Y intercept
_____
Positive
Negative

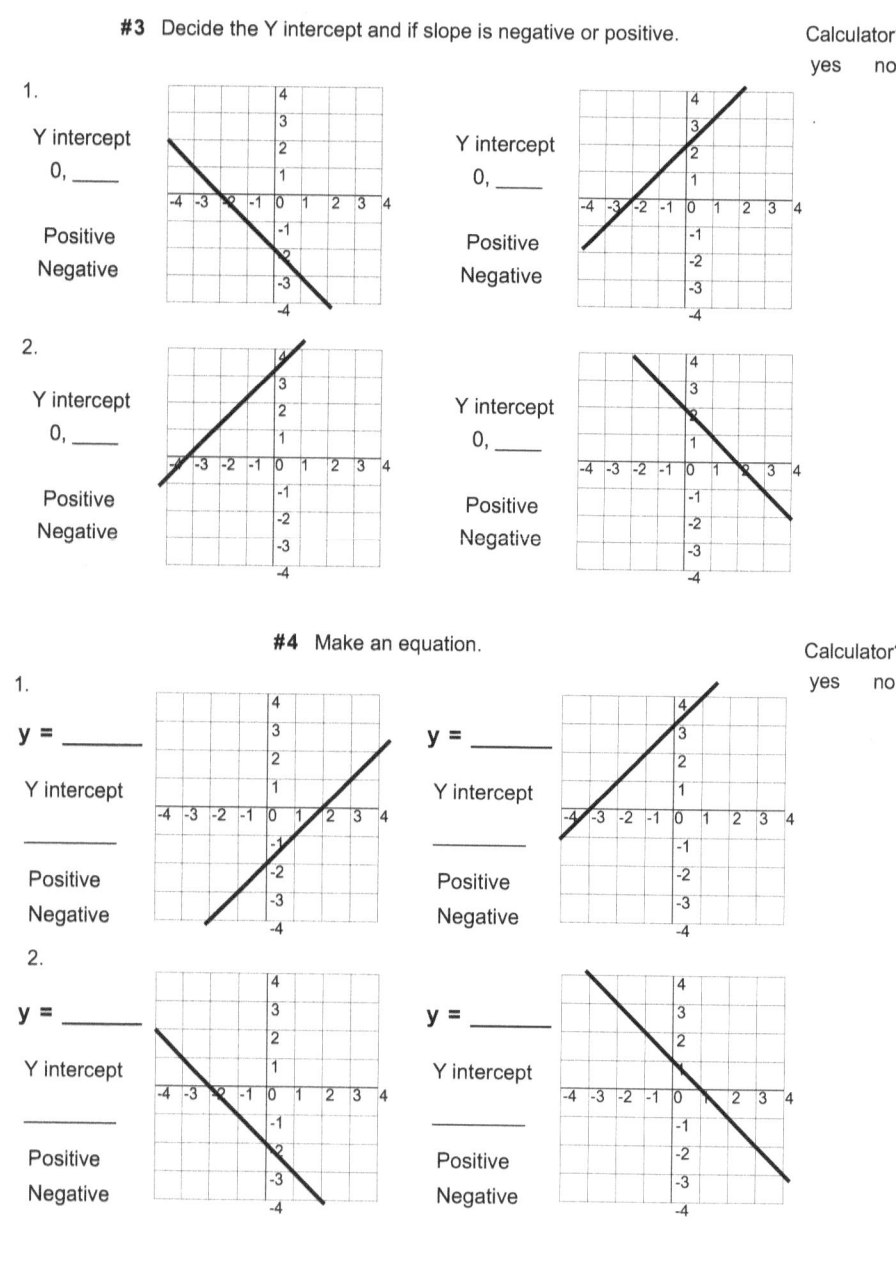

Ch 2 Ls 1 Use slope with equations. 11

_____ #1 #2 ____ / 9   #3 ____ / 4   R ___ / 9 T ____ / 22 _____
          Name   Checker

#1 1. What words show how a fraction makes slope? _____
      2. What do you look at 1st to find slope? _____
      3. When is a slope flatter than standard? _____
      4. When is a slope steeper than standard? _____
      5. Does X or Y change a slope to be negative? _____

   #2   1. Find if the slope is flatter or
            steeper. How can you tell? $\dfrac{1}{4}$

      The fraction is _____   Flatter   Steeper

         2. Is the slope flatter or steeper? $\dfrac{5}{3}$

      The fraction is _____   Flatter   Steeper

         3. What's the slope? Is it flatter,
            equal, or steeper?

                      Flatter   Steeper   Equal   [___]

         4. What's the slope?

                      Flatter   Steeper   Equal   [___]

#3 Decide the slope and circle if it's flatter or steeper.   Calculator? yes no

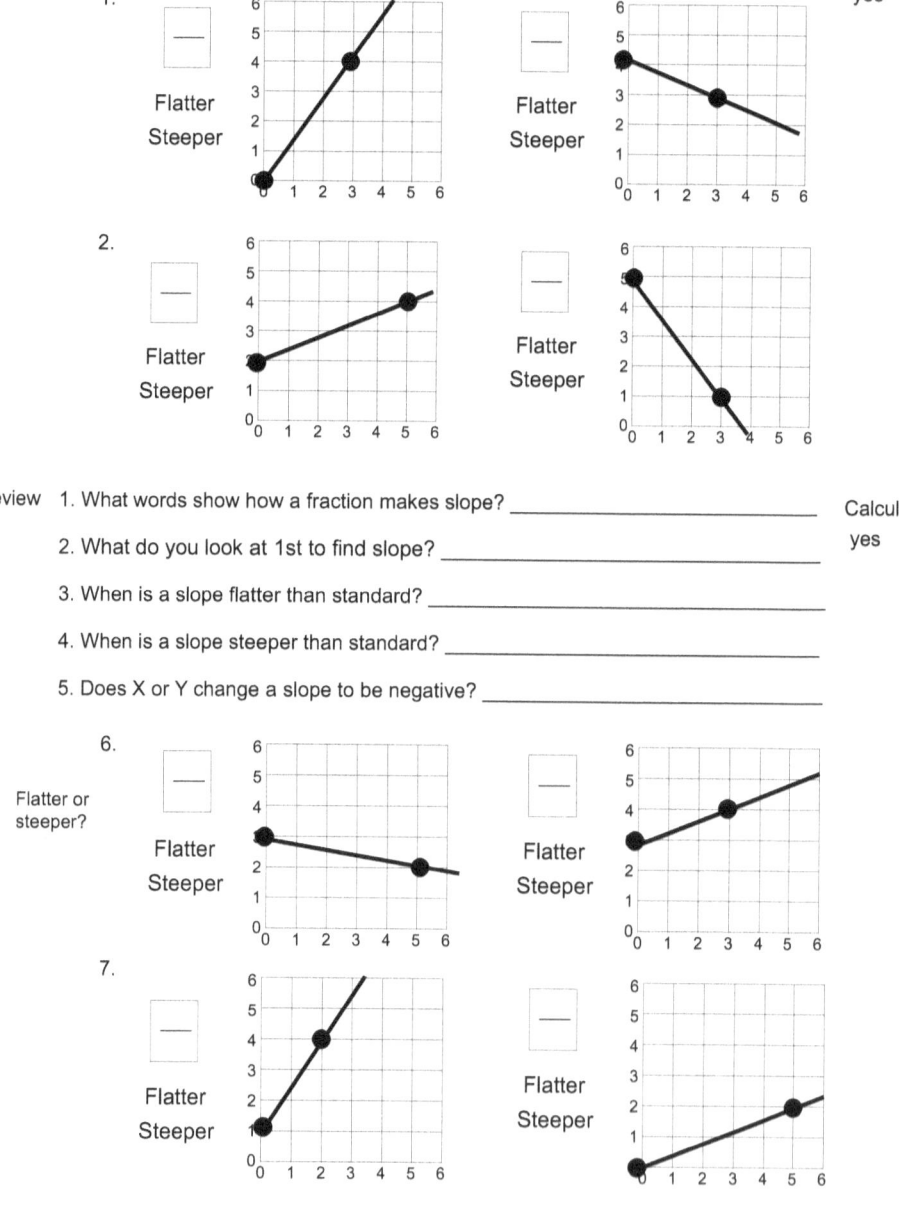

1. Flatter / Steeper   Flatter / Steeper

2. Flatter / Steeper   Flatter / Steeper

Review
1. What words show how a fraction makes slope? _____   Calculator? yes no
2. What do you look at 1st to find slope? _____
3. When is a slope flatter than standard? _____
4. When is a slope steeper than standard? _____
5. Does X or Y change a slope to be negative? _____

6. Flatter or steeper?   Flatter / Steeper   Flatter / Steeper

7. Flatter / Steeper   Flatter / Steeper

Ch 2 Ls 2 How to use slope intercept formula.   13

_____   #1 #2 ____ /7   #3 ____ /4   R ____ /7   T ____ /18 _____
                Name   Checker

#1   1. What formula puts slope together with Y intercept? _____
     2. What is the name for this formula? _____
     3. What is a standard equation? _____
     4. y = 2x + 1 How do you say this equation? _____
     _____

#2   1. What 2 things are happening?   y = 5x + 2

            What does the graph look like?   Run: ____ Rise: ____ Y int ____

                        Negative
                        Positive

     2. What 2 things are happening? $y = \frac{1}{4} x + 3$

            Predict and make the graph.   Run: ____ Rise: ____ Y int ____

                        Negative
                        Positive

     3. What 2 things are happening?   y = - 0.75x + 1

                  Negative   Positive   Run: ____ Rise: ____ Y int ____

#3 Find x is 2, then make the graph.   Calculator? yes no

1.
y = 5x + 2

_____

_____

X is 2, y is ___

y = $\frac{1}{4}$x - 5

_____

_____

X is 2, y is ___

2.
y = -x + 3.5

_____

_____

X is 2, y is ___

y = -$\frac{5}{2}$x + 1

_____

_____

X is 2, y is ___

Review 1. What formula puts slope together with Y intercept? _____   Calculator? yes no

2. What is the name for this formula? _____

3. What is a standard equation? _____

4. y = 2x + 1 How do you say this equation? _____

_____

Use Slope and Y intercept to make each equation.

5.   y = _____   y = _____   y = _____

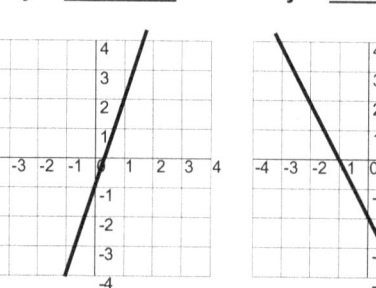
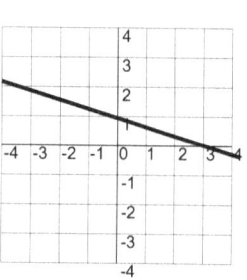

Ch 2 Ls 3 Find slope from a graph.   15

_____ #1 #2 ____/ 7   #3 ____/ 8   R ____/ 4   T ____/ 19 _____
            Name   Checker

#1 1. Where does a line start? _____
    2. What is a whole point? _____
    3. If Y intercept is not 0, what finds the slope? _____
    4. How do you say an equation? _____

#2 1. Find a whole point. What's the slope?

               Whole Point is (___, ___)   Slope is _____.

    2. Find a whole point. What's the slope?

               Whole Point is (___, ___)   Slope is _____.

    3. Find a whole point. What's the slope?

               Whole Point is (___, ___)   Slope is _____.

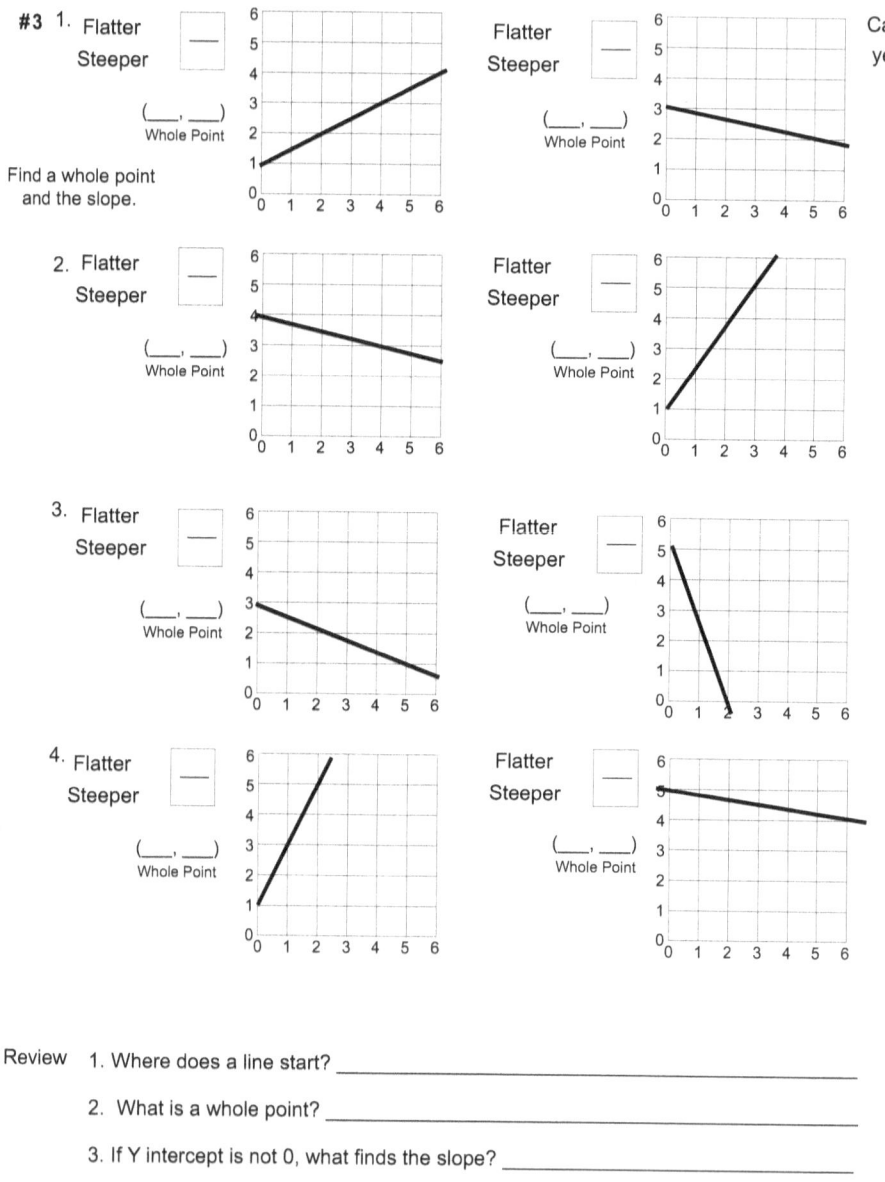

Review 1. Where does a line start? _____

2. What is a whole point? _____

3. If Y intercept is not 0, what finds the slope? _____

4. How do you say an equation? _____

Ch 2 Ls 4 How Slope Formula works. 17

_____ #1 #2 ____/ 10 #3 ____/ 6 R ____/ 10 T ____/ 26 _____
             Name   Checker

#1 1. What is the Slope Formula? _____
   2. When does the 2nd Y make a positive slope? _____
   3. When does the 2nd Y make a negative slope? _____
   4. Think of a graph. What part makes a negative slope? _____

   5. Look at the Ys. Is the
      slope positive or negative?   (0, 4) (4, 2)

                                                Positive   Negative

   6. Look at the Ys. Is the
      slope positive or negative?   (0, 8) (3, 9)

                                                Positive   Negative

#2 1. What are the Ys and Xs?   (0, 7) (8, 4)
       Find the slope.

       Xs ____ - ____ = ____   Ys ____ - ____ = ____   [—]

   2. What are the Ys and Xs?   (2, 5) (4, 6)
      Find the slope.

       Xs ____ - ____ = ____   Ys ____ - ____ = ____   [—]

   3. What are the Ys and Xs?   (3, 5) (6, 7)
      Find the slope.

       Xs ____ - ____ = ____   Ys ____ - ____ = ____   [—]

   4. What are the Ys and Xs?   (1, 4) (8, 5)
      Find the slope.

       Xs ____ - ____ = ____   Ys ____ - ____ = ____   [—]

#3 Look at the Ys. Is the slope positive or negative?   Calculator?
                                    yes no

1.   (0, 4)   (4, 2)   (0, 8)   (3, 9)
   Positive   Negative   Positive   Negative

2.   (0, - 1)   (5, - 3)   (0, 3)   (4, 2)
   Positive   Negative   Positive   Negative

3.   (0, 2)   (3, 5)   (0, - 2)   (4, - 5)
   Positive   Negative   Positive   Negative

Review 1. What is the Slope Formula? _____   Calculator?
2. When does the 2nd Y make a positive slope? _____   yes no
3. When does the 2nd Y make a negative slope? _____
4. Think of a graph. What part makes a negative slope? _____

What are the Ys? Find the Xs. Find each slope.

5.   (0, 7)   (8, 4)   Xs ___ - ___ = ___   Ys ___ - ___ = ___   ▭

6.   (1, - 2)   (5, 3)   Xs ___ - ___ = ___   Ys ___ - ___ = ___   ▭

7.   (3, - 3)   (6, - 4)   Xs ___ - ___ = ___   Ys ___ - ___ = ___   ▭

8.   (2, 3)   (7, 1)   Xs ___ - ___ = ___   Ys ___ - ___ = ___   ▭

9.   (0, - 5)   (6, 3)   Xs ___ - ___ = ___   Ys ___ - ___ = ___   ▭

10.   (0, 2)   (5, - 1)   Xs ___ - ___ = ___   Ys ___ - ___ = ___   ▭

Ch 2 Ls 5 Use negatives with slope formula. 19

_____ #1 #2 ____/ 9   #3 ____/ 6   R ____/ 9   T ____/ 24 _____
             Name   Checker

#1 1. What is the 2nd and 1st X in (- 2, 3) (1, 0)? _____
    2. How do Xs on both sides of the Y axis make a point? _____
    3. How can you tell the Ys make a negative slope? _____

    4. Look at the Ys. Is the slope   (- 1, 3)   (5, 4)
       positive or negative?

                                             Positive   Negative

    5. Look at the Ys. Is the slope   (- 2, 2)   (3, - 1)
       positive or negative?

                                             Positive   Negative

#2 1. What are the Ys and Xs?   (- 1, 5)   (4, 3)
       Find the slope.

        Xs ____ - ____ = ____   Ys ____ - ____ = ____   ▭

    2. What are the Ys and Xs?   (- 5, 7)   (5, 6)
       Find the slope.

        Xs ____ - ____ = ____   Ys ____ - ____ = ____   ▭

    3. What are the Ys and Xs?   (- 3, 1)   (2, 4)
       Find the slope.

        Xs ____ - ____ = ____   Ys ____ - ____ = ____   ▭

    4. What are the Ys and Xs?   (- 4, - 2) (- 1, - 3)
       Find the slope.

        Xs ____ - ____ = ____   Ys ____ - ____ = ____   ▭

#3 Look at the Ys. Is the slope positive or negative?   Calculator?
yes   no

1. (-1, 3) (3, 5)   (-3, 0) (3, -1)
   Positive Negative   Positive Negative

2. (-3, 4) (2, 1)   (-5, 8) (-1, 9)
   Positive Negative   Positive Negative

3. (-2, 4) (4, -1)   (-2, -1) (3, -2)
   Positive Negative   Positive Negative

Review 1. What is the 2nd and 1st X in (-2, 3) (1, 0)? _____   Calculator?
2. How do Xs on both sides of the Y axis make a point? _____   yes   no
3. How can you tell the Ys make a negative slope? _____

What are the Ys? Find the Xs. Find each slope.

4. (-2, 7) (6, 3)   Xs ___ - ___ = ___   Ys ___ - ___ = ___ |—|

5. (-5, 4) (8, 3)   Xs ___ - ___ = ___   Ys ___ - ___ = ___ |—|

6. (-3, -1) (2, 7)   Xs ___ - ___ = ___   Ys ___ - ___ = ___ |—|

7. (-4, -2) (1, -3)   Xs ___ - ___ = ___   Ys ___ - ___ = ___ |—|

8. (-1, 5) (6, -3)   Xs ___ - ___ = ___   Ys ___ - ___ = ___ |—|

9. (-6, -3) (7, -4)   Xs ___ - ___ = ___   Ys ___ - ___ = ___ |—|

Problem Review 21

_____ #1 #2 #3 ____/ 12 #4 #5 ____/ 12 Total ____/ 24
          Name

#1   1. Slope _____
     2. Slope Intercept Formula _____
     3. Slope Formula _____
     4. 4 Quadrants _____

#2. Find x is 2, then make the graph   Calculator? yes   no

1. $y = 2x + 1$

   _____
   _____
   X is 2, y is ___

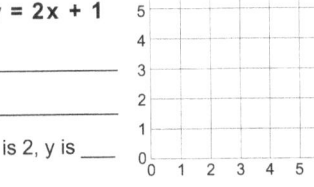

$y = \frac{3}{2}x + 1$

_____
_____
X is 2, y is ___

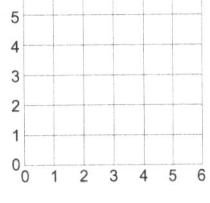

2. $y = -x + 2.5$

   _____
   _____
   X is 2, y is ___

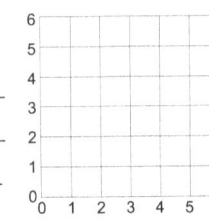

$y = -\frac{1}{4}x + 2$

_____
_____
X is 2, y is ___

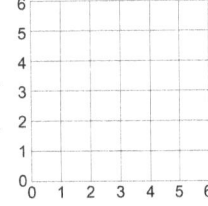

#3 Find a whole point and the slope.   Calculator? yes   no

1. Flatter ___
   Steeper

   (___, ___)
   Whole Point

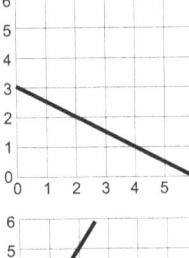

Flatter ___
Steeper

(___, ___)
Whole Point

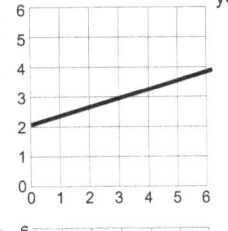

2. Flatter ___
   Steeper

   (___, ___)
   Whole Point

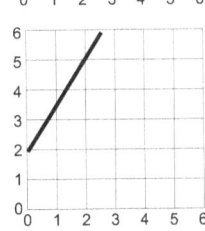

Flatter ___
Steeper

(___, ___)
Whole Point

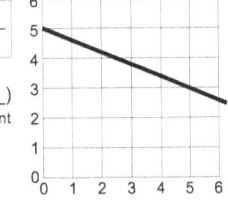

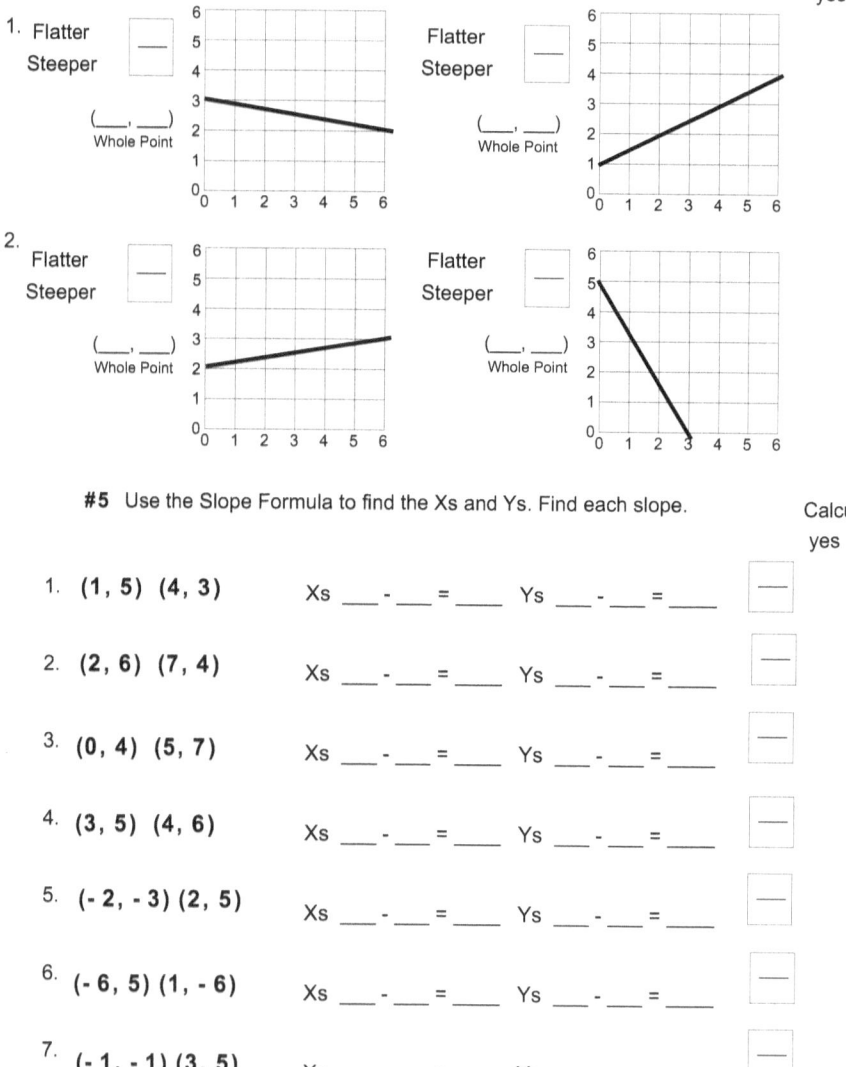

#4 Find a whole point and the slope.   Calculator? yes no

1. Flatter ☐ Steeper ☐   (__, __) Whole Point

   Flatter ☐ Steeper ☐   (__, __) Whole Point

2. Flatter ☐ Steeper ☐   (__, __) Whole Point

   Flatter ☐ Steeper ☐   (__, __) Whole Point

#5 Use the Slope Formula to find the Xs and Ys. Find each slope.   Calculator? yes no

1. (1, 5) (4, 3)   Xs ___ - ___ = ___   Ys ___ - ___ = ___   ☐

2. (2, 6) (7, 4)   Xs ___ - ___ = ___   Ys ___ - ___ = ___   ☐

3. (0, 4) (5, 7)   Xs ___ - ___ = ___   Ys ___ - ___ = ___   ☐

4. (3, 5) (4, 6)   Xs ___ - ___ = ___   Ys ___ - ___ = ___   ☐

5. (-2, -3) (2, 5)   Xs ___ - ___ = ___   Ys ___ - ___ = ___   ☐

6. (-6, 5) (1, -6)   Xs ___ - ___ = ___   Ys ___ - ___ = ___   ☐

7. (-1, -1) (3, 5)   Xs ___ - ___ = ___   Ys ___ - ___ = ___   ☐

8. (-3, -4) (2, 4)   Xs ___ - ___ = ___   Ys ___ - ___ = ___   ☐

Ch 3 Ls 1 Point on a line and midpoint of a line.   23

_____ #1 #2 ____/9 #3 ____/6 R ____/9 T ____/ 24 _____
                Name   Checker

#1 1. How do you find if a point is on a line? _____

  2. If the point is on the line, how will it solve? _____

  3. Put X in the equation and multiply.   Is 5, 4 on $y = 2x + 1$?

    Is it equal or not?   _____

    yes   no
                      _____

  4. Put X in the equation and multiply.   Is 3, 7 on $y = 2x - 1$?

    Is it equal or not?   _____

    yes   no
                      _____

#2 Find the midpoint of a line segment.

  1. What does it mean by **midpoint of a line segment**? _____

  2. How do you find the midpoint of 2 points? _____

  3. Find the Xs of this midpoint.   (- 8, 2) (2, 5)

       Find the Ys.
       What's the midpoint?   ___ ÷ 2 = ___

                                 ___ ÷ 2 = ___   Midpoint: _____

  4. What's the midpoint?   (1, 2) (5, 6)

       ___ ÷ 2 = ___   ___ ÷ 2 = ___   Midpoint: _____

  5. What's the midpoint?   (- 4, - 1) (4, 2)

       ___ ÷ 2 = ___   ___ ÷ 2 = ___   Midpoint: _____

#3 Find if the point is on the line or not.   Calculator?
                                                                              yes   no

1.   Is 2, 4 on y = 2x + 1?   Is 3, 7 on y = 3x - 2?
         ___ = 2___ + 1   ___ = 3___ - 2
     yes   no   ___ = ___   yes   no   ___ = ___

2.   Is 4, 9 on y = 2x + 1?   Is -1, -3 on y = -4x + 2?
         ___ = 2___ + 1   ___ = -4___ + 2
     yes   no   ___ = ___   yes   no   ___ = ___

3.   Is -3, -6 on y = 2x + 3?   Is 2, 7 on y = -2x + 3?
         ___ = 2___ + 3   ___ = -2___ + 3
     yes   no   ___ = ___   yes   no   ___ = ___

Review 1. How do you find if a point is on a line? _____   Calculator?
       2. If the point is on the line, how will it solve? _____   yes   no
       3. What does it mean by **midpoint of a line segment**? _____
       4. How do you find the midpoint of 2 points? _____

                    Find the midpoint of each problem.

5.   (1, 2) (5, 4)   ___ ÷ 2 = ___   ___ ÷ 2 = ___   Midpoint: ___, ___

6.   (-5, 1) (2, -3)   ___ ÷ 2 = ___   ___ ÷ 2 = ___   Midpoint: ___, ___

7.   (-7, 3) (3, 7)   ___ ÷ 2 = ___   ___ ÷ 2 = ___   Midpoint: ___, ___

8.   (-2, 4) (4, 6)   ___ ÷ 2 = ___   ___ ÷ 2 = ___   Midpoint: ___, ___

9.   (-8, 2) (2, -1)   ___ ÷ 2 = ___   ___ ÷ 2 = ___   Midpoint: ___, ___

Ch 3 Ls 2 Use Distance Formula with 2 points. 25

_____ #1 #2 ____ /7 #3 ____ /6 R ___ /9 T ____ /22 _____
           Name   Checker

#1 1. What's the 1st step to find distance between 2 points? _____
    2. What are the next 2 steps to find distance? _____
    3. What's the last step to find distance? _____
    4. Subtract Xs and Ys, what does SSS mean? _____

#2 1. Find the distance   (0, 2) (5, 4)   What's the 1st step
       from 0, 2 to 5, 4.   to find distance?

                Xs ___ - ___ = ___   Ys ___ - ___ = ___   What next in SSS?

                X squared = ___ + Y squared is ___ = ___   What last in SsS?

                Square root of ___ is about ___

    2. Find the distance   (- 3, 1) (4, 6)   What's the 1st step
       from - 3, 1 to 4, 6.   to find distance?

                Xs ___ - ___ = ___   Ys ___ - ___ = ___   What next in SSS?

                X squared is ___ + Y squared is ___ = ___   What last in SsS?

                Square root of ___ is about ___

    3. Find the distance   (- 2, - 3) (7, 2)   What's the 1st step
       from - 2, - 3 to 7, 2.   to find distance?

                Xs ___ - ___ = ___   Ys ___ - ___ = ___   What next in SSS?

                X squared is ___ + Y squared is ___ = ___   What last in SsS?

                Square root of ___ is about ___

26.

#3 Subtract Xs and Ys, then find the next 2 steps.   Calculator?
                                                         yes   no

1. (-1, 2) (5, 3)   Xs ___ - ___ = ___   Ys ___ - ___ = ___
                       X squared is ___ + Y squared is ___ = ___
                       Square root of ___ is about ___

2. (0, 2) (6, 4)   Xs ___ - ___ = ___   Ys ___ - ___ = ___
                       X squared is ___ + Y squared is ___ = ___
                       Square root of ___ is about ___

3. (-3, 1) (3, 8)   Xs ___ - ___ = ___   Ys ___ - ___ = ___
                       X squared is ___ + Y squared is ___ = ___
                       Square root of ___ is about ___

Review 1. What's the 1st step to find distance between 2 points? _____   Calculator?
       2. What's the next step to find distance? _____   yes   no
       3. What are the last 2 steps to find distance? _____
       4. Subtract Xs and Ys, what does SSS mean? _____

                       Find the distance between these points.

5. (-2, 1) (5, 4)   Xs ___ - ___ = ___   Ys ___ - ___ = ___
                       Square root of ___ is about ___

6. (1, 3) (6, 2)   Xs ___ - ___ = ___   Ys ___ - ___ = ___
                       Square root of ___ is about ___

7. (-5, 1) (5, 6)   Xs ___ - ___ = ___   Ys ___ - ___ = ___
                       Square root of ___ is about ___

Ch 3 Ls 3 Use 2 of 3 rule to find y intercept. 27

_____ #1 #2 ___ / 7 #3 ___ / 6 R ___ / 8 T ___ / 21 _____
              Name   Checker

#1 1. Name 3 things that make up 2 out of 3 rule. _____

_____

2. What does 2 out of 3 rule do? _____
3. What formula does the 2 out of 3 rule use? _____
4. You know slope and a point. What does it find? _____

#2 1. Solve in 2 steps to find the y intercept. **Slope 2 Point 3, 1**   What numbers go in the equation?

___ = ___ (__) + b   Solve the equation. What's the new B?

_____

_____

2. Use 2 steps to find the y intercept. **Slope $\frac{1}{2}$ Point 2, 3**   What numbers go in the equation?

___ = ___ (__) + b   Solve the equation. What's the new B?

_____

_____

3. Use 2 steps to find the y intercept. **Slope - 1 Point 1, 4**   What numbers go in the equation?

___ = ___ (__) + b   Solve the equation. What's the new B?

_____

What's the new parallel equation?

_____

#3 Find the new Y intercept.   Calculator? yes no

1. **Slope 4   Point 2, 1**   ___ = ___ (___) + b   New Y is ___, ___
   ___ = b

2. **Slope $\frac{1}{2}$ Point 1, 4**   ___ = ___ (___) + b   New Y is ___, ___
   ___ = b

3. **Slope 3   Point 5, 2**   ___ = ___ (___) + b   New Y is ___, ___
   ___ = b

Review 1. Name 3 things that make up 2 out of 3 rule. _____   Calculator? yes no

2. What does 2 out of 3 rule do? _____

3. What formula does the 2 out of 3 rule use? _____

4. You know slope and a point. What does it find? _____

5. How do you find the new Y intercept? $y = -3x + 1$ with 4, 2

Find a new parallel line equation.
   ___ = ___ (___) + b
   ___ = b
                                              _____
                                              Parallel Equation

6. Start with the new Y intercept. $y = 0.2x + 6$ with 3, 5
   ___ = ___ (___) + b
   ___ = b
                                              _____
                                              Parallel Equation

7. Find the Y intercept and equation. $y = -x + 4$ with 2, 2
   ___ = ___ (___) + b
   ___ = b
                                              _____
                                              Parallel Equation

8. Find the Y intercept and equation. $y = 2x + 3$ with 1, 4
   ___ = ___ (___) + b
   ___ = b
                                              _____
                                              Parallel Equation

Ch 3 Ls 4 Make a new perpendicular equation.   29

_____ #1 #2 ____ / 9   #3 ____ / 6   R ____ / 9   T ____ / 24 _____
         Name   Checker

#1 1. Name 2 steps to make a perpenduclar slope. _____ _____
    2. What is the 2 out of 3 Rule? _____
    3. y = - 2x + b with (1, 2) What's the 1st step? _____
    4. What's the 2nd step? _____
    5. What's the perpendicular equation? _____

    #2 1. What's the 1st step to perpendicular slope?   3

        What's the 2nd step to perpendicular slope?   _____

                                                            _____

        2. What's the 1st step to perpendicular slope? $\frac{1}{4}$

        What's the 2nd step to perpendicular slope?   _____

                                                            _____

        3. What's the perpendicular slope for - 2?   - 2

                                                            _____

        4. Make a perpendicular slope.   Slope $\frac{1}{2}$ Point 1, 2

              Use slope intercept.   Negative reciprocal is _____
                  1st step?   _____ = _____ (_____) + b

              Solve for B.   _____

              Make the equation.   _____

                                           _____

#3 Find the new Y intercept and perpendicular slope.   Calculator?
                                                         yes   no

1. **Slope - 2   Point 4, 1**   ___ = ___ (___) + b   New Y is ___, ___
   Perpendicular   ___ = b
   slope is ___

2. **Slope $\frac{1}{3}$ Point 2, 3**   ___ = ___ (___) + b   New Y is ___, ___
   Perpendicular   ___ = b
   slope is ___

3. **Slope - 3   Point 5, 2**   ___ = ___ (___) + b   New Y is ___, ___
   Perpendicular   ___ = b
   slope is ___

Review 1. Name 2 steps to make a perpenduclar slope. ___ ___   Calculator?
2. What is the 2 out of 3 Rule? ___   yes   no
3. y = - 2x + b with (1, 2) What's the 1st step? ___
4. What's the 2nd step? ___
5. What's the perpendicular equation? ___

6. Perpendicular slope is ___   y = 2x + 1 with 3, 2

Find a new perpen-   ___ = ___ (___) + b
dicular line equation.
                        ___ = b
                                         _____
                                         Perpendicular Equation

7. Perpendicular slope is ___   y = $\frac{2}{3}$x + 1 with 1, 5

                        ___ = ___ (___) + b
                        ___ = b
                                         _____
                                         Perpendicular Equation

8. Perpendicular slope is ___   y = 3x + 1 with 1, 4

                        ___ = ___ (___) + b
                        ___ = b
                                         _____
                                         Perpendicular Equation

Review Problem 31

_____ #1 to #4 _____ / 16   #5 to #7 _____ / 8   T _____ / 24
　　　　Name

1. Point on a Line _____
2. Midpoint of a Line Segment _____
3. Distance Formula _____
4. 2 out of 3 Rule _____
5. Parallel Line _____
6. Perpendicular Line _____

#2. 1. Is 3, 2 on y = -x + 1?   Is 2, 7 on y = 4x - 1?   Calculator?
                                                                          yes   no
Is the point   ___ = -___ + 1   ___ = 4___ - 1
on this line?   yes   no   ___ = ___   yes   no   ___ = ___

      2. Is 4, 6 on y = 2x - 1?   Is -5, 13 on y = -3x - 2?

              ___ = 2___ -   ___ = -3___ - 2
          yes   no   ___ = ___   yes   no   ___ = ___

#3 1. (1, 3)   (8, 5)   ___÷2 = ___   ___÷2 = ___   Midpoint: ___ , ___   Calculator?
                                                                              yes   no
Find the   2. (-4, 1) (6, -2)   ___÷2 = ___   ___÷2 = ___   Midpoint: ___ , ___
midpoint.
            3. (-8, 3) (5, 6)   ___÷2 = ___   ___÷2 = ___   Midpoint: ___ , ___

#4 1. (-4, 3) (2, 1)   Xs ___ - ___ = ___   Ys ___ - ___ = ___   Calculator?
                           X squared is ___ + Y squared is ___ = ___   yes   no
Find the distance
between the 2 points.   Square root of ___ is about ___

     2. (0, 3) (5, 2)   Xs ___ - ___ = ___   Ys ___ - ___ = ___
                           X squared is ___ + Y squared is ___ = ___
                           Square root of ___ is about ___

     3. (-3, 2) (5, 5)   Xs ___ - ___ = ___   Ys ___ - ___ = ___
                           X squared is ___ + Y squared is ___ = ___
                           Square root of ___ is about ___

#5 Find the new Y intercept.   Calculator?
                                                            yes   no

1. Slope $\frac{1}{3}$ Point -2, 3   ___ = ___ (___) + b   New Y is ___, ___
                                          ___ = b

2. Slope 3   Point 4, 1   ___ = ___ (___) + b   New Y is ___, ___
                             ___ = b

3. Slope 4   Point 1, 5   ___ = ___ (___) + b   New Y is ___, ___
                             ___ = b

#6 Find a new parallel line equation.   Calculator?
                                                            yes   no

1. How do you find the new Y intercept?   y = -4x - 1 with 5, 1

   ___ = ___ (___) + b
   ___ = b
                                            _____
                                              Parallel Equation

2. Start with the new Y intercept.   y = 0.5x + 2 with 4, 2

   ___ = ___ (___) + b
   ___ = b
                                            _____
                                              Parallel Equation

#7 Find the new Y intercept and perpendicular slope.   Calculator?
                                                            yes   no

1. Slope -3   Point 4, 1   ___ = ___ (___) + b   New Y is ___, ___
   Perpendicular   ___ = b
   slope is _____

2. Slope $\frac{1}{4}$ Point 2, 3   ___ = ___ (___) + b   New Y is ___, ___
   Perpendicular   ___ = b
   slope is _____

3. Slope -2   Point 3, 4   ___ = ___ (___) + b   New Y is ___, ___
   Perpendicular   ___ = b
   slope is _____

Ch 4 Ls 1 How 2 variables change equations. 33

_____ #1 #2 ____/ 6   #3 #4 ____/ 4   R ___/ 6   T ____/ 16 _____
         Name   Checker

#1 1. How does the answer change when a 1 variable uses 2? _____
    2. **15 x 2 = t** Total is what kind of variable? _____
    3. **k x 2 = t** Kids is what kind of variable? _____
    4. Which variable goes at the bottom of a graph? _____

#2 1. Make a 1st equation.   Mr D drove 80 kph for 3 hours.
                                    How far does he drive?

       Change for driving some hours.   _____

       What does the graph look like?   _____

                                         0 1 2 3 4 5 6 7 8 9 10

                                         _____

    2. Make a 1st equation.   Amav saves R 300 each week for 10 weeks.
                                    How much did he save?

       Change for saving some money.   _____

       What does the graph look like?   _____

                                         0 1 2 3 4 5 6 7 8 9 10

                                         _____

#3 1. Mitlul bought 4 shirts that are Rs 370 each. Change the equation if he doesn't know how many to get.

Calculator? yes   no

_____

_____

_____

Graph it

0 1 2 3 4 5 6 7 8 9 10

_____

2. Divit drove 60 kph for 4 hours. Change the equation if he feels like driving farther.

_____

_____

_____

Graph it

0 1 2 3 4 5 6 7 8 9 10

_____

Review 1. How does the answer change when a 1 variable uses 2? _____

Calculator? yes   no

2. **15 x 2 = t** Total is what kind of variable? _____

3. **k x 2 = t** Kids is what kind of variable? _____

4. Which variable goes at the bottom of a graph? _____

5. KPH Hrs Kilometers

How does the 2nd variable change what's happening.

60 • 3 = d

60 • x = y

Problem about TJ's trip.

_____

_____

6. Rs/hr hours total

12 • 4 = d

12 • x = y

Problem about Ojas's work

_____

_____

Ch 4 Ls 2 Use y intercepts with 2 variables. 35

_____ #1 #2 ____ / 5   #3 ____ / 4   R ___ / 5   T ____ / 14 _____
        Name   Checker

#1 1. How does a different Y intercept change a graph? _____
    2. What equation works with both slope and Y intercept? _____
    3. Does slope or Y intercept use rate formula? _____

#2 1. Make a 2 variable equation.   Dr T drove 70 kph for some hours.
                                        How far does he drive?

        Same equation, but 40 mph.

        Make a graph for both lines.

    2. Make a 2 variable equation.   TJ does 2 more chin ups than Ojas.
                                        Ojas wonders how many he can do?

        Change for 2 more chin ups than Bob.

        Make a graph for both lines.

#3

1. TJ bought 3 pants that are Rs 450 each. Change the equation if he doesn't know how many to get. _____

Calculator? yes   no

Solve these equations and graph it.

_____

_____ Graph it

0 1 2 3 4 5 6 7 8 9 10

_____

2. Adah made Rs 50/hr for 4 hours. Change the equation if she doesn't know how many hours she'll get. _____

_____

_____

_____ Graph it

0 10 20 30 40 50

_____

Review **1.** How does a different Y intercept change a graph? _____

Calculator? yes   no

2. What equation works with both slope and Y intercept? _____

3. Does slope or Y intercept use rate formula? _____

Describe each equation.

4.
```
soccer  extra  total
balls
  x  +  5  =  n
  8  +  5  =  n
Amav has team soccer balls.
```

_____

_____

5.
```
rate  time    base  total
 75  (H)   +   80  =  t
 75  (10)  +   80  =  t
```
This is what Ojas makes for work. How much money is it?

_____

_____

Ch 4 Ls 3 Predict results using 2 variables. 37

_____ #1 #2 ____/ 7   #3 ____/ 4   R ___/ 7 T____/ 18 _____
             Name   Checker

#1 1. In this problem, what part of the equation is R40 a day? _____
  2. How did it write 10% of the R200 he sold? _____
  3. How does it change for whatever he sells? _____
  4. What do prediction equations do? _____
  5. What part of the equation starts the prediction? _____

#2 1. What's the formula?   Mr K makes Rs 50 a day plus comission. His commission is 10% of Rs 64,000. What's the equation?

Make a graph for it.   _____

(graph with x-axis: 0 200 400 600 800 1000)

2. What's the formula?   Eva spent Rs 30 on 2 packs of chips, then bought some pops at Rs 40 each. Find the equation.

What's the 1st step?   _____

Make a graph for it.   _____

(graph with x-axis: 0 1 2 3 4 5 6 7 8 9 10)

#3 1. Mr K makes Rs 100 a day plus 20% comission. How much does he make for 1 day? Start at 0 and go to 2500.

Predict results using 2 variables.

_____

Calculator?   yes   no

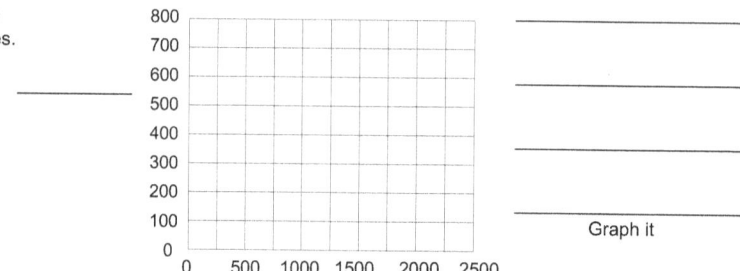

_____
_____
_____
_____
Graph it
_____

2. TJ makes Rs 800 a week from his 1st job. He gets a 2nd job at Rs 100/hr. He can work upto 30 hrs a week. Solve it.

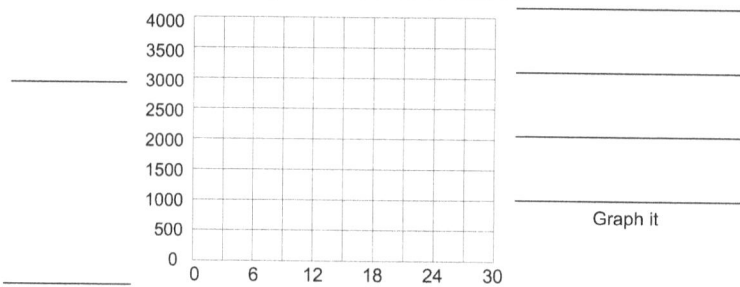

_____
_____
_____
_____
Graph it
_____

Review 1. In this problem, what part of the equation is Rs 100 a day? _____

Calculator?   yes   no

2. How did it write 10% of the Rs 3000 he sold? _____

3. How does it change for whatever he sells? _____

4. What do prediction equations do? _____

5. What part of the equation starts the prediction? _____

6.   R/hr Hrs Tips total

Solve it.
$35(H) + X = d$
$35(6) + 28 = d$
How Amav makes money.

_____
_____

7.   % sales base total
$25\% \ s + 400 = t$
$5\%(R300,000) + 400 = t$
How Mr J, a realtor, gets paid.

_____
_____

Ch 4 Ls 4 How Add Rate Formulas use 2 variables.   39

_____ #1 #2 ___/7 #3 ___/4 R ___/5 T ___/16 _____
              Name   Checker

#1 1. How can Add Rate Formulas use a 2nd variable? _____
        2. What happens when you solve the 1st step for Add Rates? _____
        _____
        3. How many things can it solve for when Add Rates switches variables? _____

#2 1. Mr K's 1st trip is 30 kph for 1 hour and   What's the 1st equation?
        then 70 kph for 3 hrs. How far did he go?

        (___ x ___) + (___ x ___) = ___   He decides to go longer the 2nd trip.
                                               Use a 2nd variable with the equation.

        (___ x ___) + (___ x ___) = ___   Solve the 1st step.

        _____   What does the graph look like?

                                               800
                                               700
              Y intercept _____   _____   600
                                               500
              Slope _____   400
                                               300
                                               200
                                               100
                                     _____   0
                                                  0 1 2 3 4 5 6 7 8 9 10

        2. Two basketballs are Rs 390 ea. How many   What's the 1st equation?
           baseballs at Rs 120 ea can he get for Rs 1500?

        (___ x ___) + (___ x ___) = ___   He has more money so he'll buy
                                               more baseballs. Use a 2nd variable.

        (___ x ___) + (___ x ___) = ___   Solve the 1st step.

        _____   What does the graph look like?

                                               Rs1600
                                               Rs1600
              Y intercept _____   _____   Rs1200
                                               Rs1000
              Slope _____   Rs 800
                                               Rs 600
                                               Rs 400
                                     _____   Rs 200
                                                0
                                                  0 1 2 3 4 5 6 7 8 9 10

#3 1. Mr K's trip is 40 kph for 0.5 hrs and then he goes
    for 70 kph for upto 10 hours. How far did he go?

    Calculator?   yes   no

    _____

    _____

    _____

    Graph it

    0 1 2 3 4 5 6 7 8 9 10

2. Six oil filters are Rs 500 ea. How many wipers at
   Rs 250 each can he get to for the order?

   _____

   _____

   _____

   Graph it

   0 1 2 3 4 5 6 7 8 9 10

Review 1. How can Add Rate Formulas use a 2nd variable? _____

Calculator?   yes   no

2. What happens when you solve the 1st step for Add Rates? _____

_____

3. How many things can it solve for when Add Rates switches variables? _____

What is happening in each equation?

4.   KPH Hrs   KPH Hrs   KM
     $20(0.5) + 65(T) = d$ _____
     $20(0.5) + 65(3) = d$ _____
     Ojas goes on a trip.

5.   1st Job   2nd Job
     R/hr hrs   R/hr hrs   total
     $50(10) + 42(H) = t$ _____
     $50(10) + 42(20) = t$ _____
     Anvi has 2 jobs.

Problem Review 41

_____ #1 #2 ____/7 #3 ____/7 Total ____/14
Name

#1 1. 1 to 2 Variables _____
    2. Prediction Equations _____

        #2 Make an equation and change it for a 2nd variable.   Calculator?
                                                                    yes   no

    1. Mr K bought 3 pants that are Rs 700 each and a
       shirt for Rs 500.   _____

       There's a sale!!! Now the pants are Rs 600 each.   _____
       How much did he spend?
                                                                _____

    2. Ms T drove 20 kph for 0.5 hours and 80 kph for
       3 hours. How far did she go?   _____

       She decides to drive 5 hours. How far can   _____
       she go?
                                                                _____

    3. Reya bought 4 bottless of ketchup, each Rs 170.
       How much did it cost?   _____

       She changes it to 6 bottles of ketchup.   _____

                                                                _____

    4. Mitul works 6 hours that pays Rs 80 per hour.
                                                                _____
       He wants a raise. What's the new equation?
                                                                _____

                                                                _____

    5. Ojas got 3 pizzas for the futball team that were
       Rs 420 each.   _____

       Oooops!! Not enough. He'll need more. He makes   _____
       it some pizzas for Rs 420 each. Write an equation.
                                                                _____

#3 Make an equation before solving it.   Calculator? yes no

1. Mrs W makes Rs 200 a day plus 10% on what she sells. How many sales does she need to make Rs 600 for 1 day?

2. TJ makes Rs 400 a week in a part time job. He gets a 2nd job at Rs 60/hr, where he will work 30 hours a week. How much will he make in a week?

3. Mr T's trip was 20 kph for 0.3 hrs and then 90 kph for some time. Make an equation for it and solve for 4.5 hours at 90 kph. How far did he go?

4. JJ needs to spend Rs 2000 on his guests. He needs 2 cappuccinos at Rs 70 each. How many sodas can he get at Rs 30 each can he get to fill out the order?

5. A train in China goes 180 kph for 4 hours in the couintry but slows to 50 kph for a 0.8 hour in the city. How far does it go on it's trip?

6. A man drives a truck 10 km roundtrip to get to work. He then drives a semi at 80 kph for some time. Today he works 12 hours. How far does he go in a day?

7. Mr W makes Rs 80 an hour as a policeman for 8 hours. Then he goes to a store and works for R 100 an hour for some hours. He works 2 hours today. How much does he make today?

Ch 5 Ls 1   How to use x intercepts.   43

_____   #1 #2 ____ / 8   #3 ____ / 4   R ____ / 7   T ____ / 19 _____
         Name   Checker

#1   1. How do you find an X intercept from an equation? _____
     2. What formula does an X intercept find? _____
     3. Name 2 steps to find X intercepts? _____
     4. What kind of problems use an X intercept to solve it? _____
     5. What does the graph look like? _____

#2   1. What's the 1st step to an X intercept?   y = 4x + 12

         Put 0 in for the _____ .   _____

              What's the X intercept?   _____

     2. What's the X intercept?   y = 3x + 15

                                                     _____

                                                     _____

     3. Mr K has 30 kilometers to run on a hike. If   What equation uses
        he runs 12 kph, how long until he finishes?   X intercept to solve it?

                    Solve the 1st step.   _____

                  What's the X intercept?
                 What does the graph look like?   _____

                                                 0 1 2 3 4 5 6 7 8 9 10
                    _____   - 10
         Y intercept _____   - 20
                                                 - 30
             Slope _____   - 40
                    _____   - 50
                                                 - 60
                                                 - 70
                                                 - 80

#3 1. Mr B has 300 km to drive. He drives 60 kph. Calculator?
How long until he finishes? yes   no

X intercept story problems.

Total   0 1 2 3 4 5 6 7 8 9 10
- 40
- 80
kms   - 120
to go - 160
- 200
- 240
- 280
- 320

_____

_____

_____

_____
Graph it

2. Ojas has Rs 4000. He spends Rs 600 each month on his cell plan. How many months does he have enough for?

$ spent   0 1 2 3 4 5 6 7 8 9 10
- 500
- 1000
Rupees - 1500
- 2000
- 2500
- 3000
- 3500
- 4000

_____

_____

_____

_____
Graph it

Review 1. How do you find an X intercept from an equation? _____   Calculator?

2. What formula does an X intercept find? _____ yes   no

3. How can you remember how to find X intercepts? _____

4. What kind of problems use an X intercept to solve it? _____

5. What does the graph look like? _____

6.   KPH Hrs Km   time

Solve it.   80(T) - 120 = 520

How Ojas drove the 520 km trip.

_____

_____

7.   R/hr Hrs Rupees total

70(H) - 280 = y

TJ owes 280 rupees. He works 10 hours. What does he have left?

_____

_____

Ch 5 Ls 2 Both Intercept Story Problems. 45

_____ #1 #2 ____ / 6   #3 ____ / 4   R ____ / 6   T ____ / 19 _____
                Name   Checker

#1 1. What formula uses Both Intercepts to solve a problem? _____
    2. What's the 1st step to solve Both Intercepts Equation? _____
    3. What does the graph look like? _____
    4. What kind of problem would use Both Intercepts? _____

#2 1. Find both intercepts. **Mr J wants to sell Rs 80,000 of pop. Diet is Rs 2500 per case and regular is Rs 2000 per case. How much does he sell of each?**

            Solve them.   _____   _____

            Describe the graph. _____   _____

    2. Find both intercepts. **Eva wants to lose 300 calories. She'll walk for 2 cal/min and exercise for 3 cal/min. How many of each?**

            Solve them.   _____   _____

            Describe the graph. _____   _____

#3

1. A buyer has Rs 600 to buy meat for a picnic. She can buy fish for Rs 35 ea or chicken for Rs 25 ea. At what level can she best use her money?

Calculator? yes no

_____   _____   _____
_____   _____
_____   _____
_____   _____
_____   _____

2. A lawn seed comes in Rs 500 or Rs 800 form. A person comes in with Rs 20,000. Write an equation that shows how can best spend his money.

_____
_____   _____
_____   _____
_____   _____
_____   _____

Review

1. What formula uses Both Intercepts to solve a problem? _____

Calculator? yes no

2. What's the 1st step to solve Both Intercepts Equation? _____

3. What does the graph look like? _____

4. What kind of problem would use Both Intercepts? _____

Find the two levels.

5. Calories
Lifting Running total
$5x + 10y = 300$
Kate's workout.

_____
_____
_____

6. R per kilogram of Candy
R/kg lkg R/kg lkg total
$20x + 50y = 500$
Jen buys Rs 50 in candy.

_____
_____
_____

Ch 5 Ls 4 2 Equation Problems to find profit. 47

_____ #1 #2 ____/ 6 #3 ____/ 4 R ___/ 6 T ___/ 16 _____
           Name   Checker

#1 1. What's the 1st thing to make an equation? _____

    2. What's the 2nd equation. _____

    3. 8(80) + 8(20) = t How would the equation change if 20 books were sold at 25% off?
    _____

    4. How could a variable change the percent? _____

#2 1. Sell 100 books for Rs 50, and 100 books are Rs 90   Make an equation
       each. You paid Rs 7,500 to print them. Find the profit.   with the percent off.

       _____   Change to percent paid.

       _____   What's the 1st step?

       _____   Solve it for money made.

       _____   Solve for profit.

       _____

    2. Spend Rs 12,000 to print 300 CDs. Half are full price at Rs   Make an equation
       200 each and half are discounted 20% off. What's the profit?   with the percent off.

       _____   Change to percent paid.

       _____   What's the 1st step?

       _____   Solve it for money made.

       _____   Solve for profit.

       _____

48.

#3
Two Equation Problems to find profit.

1. Mr. K spent Rs 6000 to print out 300 books. 200 are full priced at Rs 80 ea but he sells 100 at Rs 60. How much will he make from them?

Calculator?
yes   no

2. Spend Rs 7000 to print 400 CDs. 3/4ths are full price at Rs 90 each and a 4th are discounted 50% off. What's the profit?

3. Mr G spent Rs 4000 to print 500 books. He sold 400 at Rs 50 each and reduced 100 of them 25%. What's his profit?

3. Zara got a book deal!!! They'll put Rs 9000 down for 500 books. She thinks they'll 300 books for Rs 60 each and it will end up as 40% for the rest of them. How much should she make?

Review
1. Where do you start to say an equation? _____
2. What's the 1st thing to look for to make an equation? _____
3. What part makes the slope? _____
4. Where is the question in an equation? _____

Calculator?
yes   no

Ch 5 Ls 4 How slope intercept uses 2 variables. 49

_____ #1 #2 ____/ 4 #3 ____/ 4 R ___/ 4 Total ____/ 12 _____
Name   Checker

#1 1. Where is the question in an equation? _____
    2. Does slope or Y intercept use rate formula? _____

#2 1. What's the equation?   Mr K makes Rs 100 a day plus he gets 20% commision
                                on what he sells. The max sold is Rs 2500.

                                              _____
                                                                          Make a graph for it.

            _____
            Find the point for R4000 sold.

2. What's the equation?   Ojas spent Rs 50 on chips, then bought some
                          pops at Rs 40 each. Find the rate formula.

                                              _____
                                                                          Make a graph for it.

            _____
            Find the point where
            he bought 4 pops..

#3

How slope intercept uses 2 var.

1. Mr B makes Rs 200 a day plus he gets 10% commision on what he sells. What does he make on 0 to Rs 5,000?

Money Mr B makes (y-axis: 0, 200, 400, 600, 800, 1000, 1200, 1400, 1600)

Money Mr B Sells (x-axis: 0, 1000, 2000, 3000, 4000, 5000)

Calculator? yes   no

_____
_____
_____
_____
_____

2. Mrs W makes Rs 400 a week plus she gets 20% commision on what she sells. What does she make upto Rs 10,000?

Money she makes (y-axis: 0, 400, 800, 1200, 1600, 2000, 2400, 2800, 3200)

Money the company makes (x-axis: 0, 2000, 4000, 6000, 8000, 10,000)

_____
_____
_____
_____

Review

1. What equation works with both slope and Y intercept? _____

2. Does slope or Y intercept use rate formula? _____

What's Happening?

3. Rs/hr hr Rs/hr hours total

   $60h + 70(30) = 3{,}300$

   How Riya works.

4. KPH hrs KM total

   $80t + 6 = d$

   How many km Ojas drives for business per day.

Calculator? yes   no

_____
_____
_____
_____

Review Problem 51

_____ #1 #2 #3 ____/ 7   #4 ____/ 7 R ____/ 9 T ____/ 23
              Name

#1 1. X Intercept _____
      2. X Intercept Problems _____
      3. Both Intercept Problems _____
      4. Profit Equations _____

#2 1. Mr K has 60 km to bike on his day off. If he bikes 10 kph, how long until he finishes?

Calculator? yes   no

X intercept story problems.

Total / Kilometers to go (0 to -80, by 10s), horizontal 0–10

Graph it

2. Ira has 290 km to drive. She drives 70 kph. How long will it take to get there?

Rs spent / Kilometers to go (0 to -320, by 40s), horizontal 0–10

Graph it

#3 1. Mitul wants to lose 200 calories at the gym. He read where weights is 4 calories per min and running is 5. How many minutes of each?

Calculator? yes   no

Both Intercept Story Problems.

Graph: 0 to 40 (by 10s) vertical, 0 to 50 (by 5s) horizontal

_____
_____

#4

1. The fee for your cell service is Rs 800 per month. Make an equation to show it.

   Solve it for 4 months.

   Calculator? yes   no

2. Mr K spent Rs 8000 to print out 200 books. 160 of them at full priced at Rs 80 each but he sells 20% at 25% off. How much will he make from them?

3. The temperature at 4 pm was 20 C. By 10 P.M., the temperature was 16 C. Find the rate of change of the temperature per hour.

4. JJ was 90 cm tall. Twenty months later, he was 100 cm tall. Find the rate of change for JJ's height.

5. The average weight of a baby at birth is 4 kg. The doctor says it should gain every 0.5 kg each week for 6 months. How much will it gain?

6. Mr B makes Rs 400 a day plus he gets 15% commision on what he sells. Make an equation..

   Solve it for Rs 12,000 for 1 day.

7. Ms G spend Rs 8000 to print 300 CDs. Half are full price at Rs 120 each and half are discounted 25% off. What's the profit if she sells tham all?

Ch 6 Ls 1 Change mph with feet per second. 53

_____ #1 #2 ____ / 7   #3 #4 ____ / 16 R ___ / 15 T ____ / 38 _____
       Name   Checker

#1 1. Why are meters per second important to problems? _____
    2. How can you change meters per second to kph? _____
          3. What's the 1st step to estimate kph?   **30 m/sec**
                    What finds the estimate?   30 has _____ 5s.
                    Solve the estimate.   ____ x ____ = ____
                                                         ____ kph
          4. What's the 1st step to estimate kph?   **40 m/sec**
                    What finds the estimate?   40 has _____ 5s.
                    Solve the estimate.   ____ x ____ = ____
                                                         ____ kph

#2 1. How can you change kph to feet per second? _____
    _____
          2. What's the 1st step to estimate ft/sec?   **20 kph**
                    What finds the estimate?   20 has _____ 10s.
                    Solve the estimate.   ____ x ____ = ____
                                                         ____ meters/sec
          3. What's the 1st step to estimate ft/sec?   **50 kph**
                    What finds the estimate?   50 has _____ 10s.
                    Solve the estimate.   ____ x ____ = ____
                                                         ____ meters/sec

#3 Change meters per second with kph. Calculator? yes no

1. 5 m/sec = _____ kph   25 m/sec = _____ kph
2. 30 m/sec = _____ kph   40 m/sec = _____ kph
3. 20 m/sec = _____ kph   80 m/sec = _____ kph
4. 35 m/sec = _____ kph   100 m/sec = _____ kph

#4 Change kph with meters per second. Calculator? yes no

1. 10 kph = _____ m/sec   40 kph = _____ m/sec
2. 70 kph = _____ m/sec   100 kph = _____ m/sec
3. 20 kph = _____ m/sec   60 kph = _____ m/sec
4. 80 kph = _____ m/sec   110 kph = _____ m/sec

Review 1. Why are meters per sec important to problems? _____ Calculator? yes no
2. How can you change meters per sec to estimate mph? _____
3. How can you change kph to meters per second? _____

4. 10 m/sec = _____ mph   15 m/sec = _____ kph
5. 70 m/sec = _____ mph   50 m/sec = _____ kph
6. 30 kph = _____ m/sec   90 kph = _____ m/sec
7. 50 kph = _____ m/sec   120 kph = _____ m/sec

8. The speed limit is 80 kph. Mrs W is going 30 m/sec. Is she under the limit? _____
_____

9. A highway has a slow limit of 60 kph. Is 20 m/sec fast enough? _____

10. A rocket is going 700 m/sec. How many kph is that? _____

11. If it was going 500 m/sec, how fast is that? _____

Ch 6 Ls 2 Use rate formula to find rate. 55

_____ #1 #2 ____/ 10 #3 ____/ 8 R ____/ 7 T ____/ 25 _____
            Name   Checker

#1 1. When does rate formula find Rate? _____
    2. What kind of graph does it use? _____
    3. What does it find on a graph? _____
    4. How does rate make a label? _____

    5. How do you solve for a rate?   r x 4 = 200
                                           seconds meters

            Divide by _____   ____ per _____

    6. How do you solve for a rate?   r x 5 = 30
                                           hours kilometers

            Divide by _____   ____ per _____

#2 1. What is the slope intercept formula? _____
    2. How does rate formula use this formula? _____
    _____

    3. What is happening in this equation?   y = 50t + 10
                                             kph kilometers

            Solve for t is 5. Travel _____ then go _____ mph.

    4. What is happening in this equation?   y = 20t + 30
                                             feet/sec meters

            Solve for t is 5. Travel _____ then go _____ ft/sec.

#3

Use slope intercept to find distance.

1. **Go 70 m, then 20 m/sec in 8 sec.**

   Solve it. (___ x ___) + ___ = ___

   Graph it. ___ = ___

   sec   0 1 2 3 4 5 6 7 8 9 10

Go 50 m, then 40 m/sec in 12 sec.   Calculator? yes no

   (___ x ___) + ___ = ___

   ___ = ___

   sec   0 1 2 3 4 5 6 7 8 9 10

2. **Go 170 m, then 25 m/sec for 5 sec.**

   Solve it. (___ x ___) + ___ = ___

   Graph it. ___ = ___

   sec   0 1 2 3 4 5 6 7 8 9 10

Go 300 m, then 40 m/sec in 20 sec.

   (___ x ___) + ___ = ___

   ___ = ___

   sec   0 1 2 3 4 5 6 7 8 9 10

Review

1. When does rate formula find Rate? _____   Calculator? yes no
2. What kind of graph does it use? _____
3. How does rate make a label? _____
4. What is the slope intercept formula? _____
5. What is the rate formula from a start Formula? _____

Solve it.

6. kph hr km total

   $80h + 10 = t$ _____

   How far BJ's car goes in 5 hr.   _____

7. m/s sec meters total

   $28s + 30 = t$ _____

   How far Mike's truck goes in 12 sec.   _____

Ch 6 Ls 3 Use percent with rate formula.   57

_____   #1 #2 ____ / 6   #3 ____ / 6   R ____ / 6   T ____ / 18 _____
                Name   Checker

#1 1. What is the slope intercept formula? _____

2. What is the rate formula from a start formula? _____

#2 1. A jet flies 500 kph for 5 hrs. It's 10% slower because   What's the expression?
of a headwind. How far will it go in the 5 hrs?

_____   Multiply 2 numbers.

_____   What's the answer?

_____   If the jet has a 2000 km trip, how much longer will it take because of a 10% off wind?

_____   What's the 1st step?

_____   What's the answer?

2. A biker rides at 20 m/sec until he goes 30% slower   What's the expression?
because of a hill. How far will he go in 10 seconds?

_____   Multiply 2 numbers.

_____   What's the answer?

_____   The biker rides a hill that's 400 m long. How much longer will it take because of the hill?

_____   What's the 1st step?

_____   What's the answer?

_____

58.

#3 1. A jet usually flies at 550 kph, but it's 4% slower because of a headwind. The jet takes 2.5 hours to fly a route. How late will it be because of the wind? _____

Calculator? yes   no

2. If a military jet has a 3 hour trip at 800 kph, how much longer will it take because of the wind that slows it down 5%? _____

3. A biker rides at 8 m/sec until he goes 7% slower because of a hill. How far will he go in 40 sec? _____

Review 1. What is the slope intercept formula? _____

Calculator? yes   no

2. What is the rate formula from a start Formula? _____

Solve each one.

3. % Hill   m/sec   sec   feet
    1.20   (5)   80 = d
    How far Phil bicycles. _____

4. % Wind   m/sec   min   feet
    0.95   (40)   20 = d
    How far Mrs G motorcycles. _____

5. % Wind   m/sec   Hrs   total
    0.90   (150)   2.5 = d
    How far Mr K flies. _____

6. % Wind   m/sec   Hrs   total
    1.10   (170)   2.4 = d
    How far the return trip takes. _____

Review Problems. 59

_____ #1 #2 #3 #4 ____ / 22 #4 #5 ____ / 6 Total ____ / 28
            Name

#1 1. Meters to KPH _____
       2. KPH to Meters per Second _____
       3. Add a Trip Equation _____
       4. Percent of Rate Formula _____

#2 Change mph with feet per second.   Calculator? yes no

1. 50 kph = ____ m/sec   30 kph = ____ m/sec
2. 70 kph = ____ m/sec   100 kph = ____ m/sec
3. 40 kph = ____ m/sec   120 kph = ____ m/sec
4. 90 kph = ____ m/sec   10 kph = ____ m/sec

#3 Change feet per second with mph.   Calculator? yes no

1. 20 m/sec = ____ kph   100 m/sec = ____ kph
2. 40 m/sec = ____ kph   30 m/sec = ____ kph
3. 10 m/sec = ____ kph   70 m/sec = ____ kph
4. 80 m/sec = ____ kph   120 m/sec = ____ kph

#4 Use Rate Formula to find Rate.   Calculator? yes no

1. Go 20 meters, then 30 m/sec in 10 sec.   Go 40 m, then 20 m/sec in 10 sec.

   Solve it. (___ x ___) + ___ = ___   (___ x ___) + ___ = ___

   Graph it.   ___ = ___   ___ = ___

Use slope intercept
to find distance.

            sec   sec
                0 1 2 3 4 5 6 7 8 9 10   0 1 2 3 4 5 6 7 8 9 10

Continued from page 59.   Calculator? yes no

2. Go 120 meters, then 25 m/sec for 10 sec.   Go 200 m, then 50 m/sec in 10 sec.

Solve it. (___ x ___) + ___ = ___   (___ x ___) + ___ = ___

Graph it.   ___ = ___   ___ = ___

sec   sec
     0 1 2 3 4 5 6 7 8 9 10   0 1 2 3 4 5 6 7 8 9 10

#5 1. Usually, a jet flies 420 kph, but it's 10% slower because of a headwind. How far will the jet go in the 3 hrs?

Calculator? yes no

Use percent with rate formula.

2. A biker rides at 30 m/sec until she goes 10% faster because of a downhill. How far will she go in 40 seconds?

3. A jet flies a 1050 mile trip at 3 hours at 450 kph. How much longer will it take because of a wind that slows it down 12%?

4. A runner runs at 8 m/sec, but he's 10% slower because of the wind. How far will he go in 120 seconds?

Ch 7 Ls 1 Find total and unit acceleration. 61

_____ #1 #2 ____/9 #3 ____/12 R ___/9 Total ____/30 _____
Name   Checker

#1 1. What does acceleration measure? _____
    2. How do you find total acceleration? _____
    3. What is unit acceleration? _____
    4. How does total acceleration find unit acceleration? _____
    5. How does it make a label for unit acceleration? _____
    6. What kind of graph does acceleration use? _____

#2 1. What is the total acceleration? **Accelerate 20 - 40 kph in 4 sec**

    What is the unit acceleration? _____

                                           ____ per _____

    2. What is the total acceleration? **Accelerate 15 - 60 m/sec in 5 sec**

    What is the unit acceleration? _____

                                           ____ per _____

3. What's the total acceleration?

                                 Meters/sec — graph from 0 to 60, line rising from (0,0) through approx (5,20) then flat to (6,20)
                                 Seconds (0 1 2 3 4 5 6)

    Total acc _____   What's the unit acceleration?

                                   ____ per _____

#3

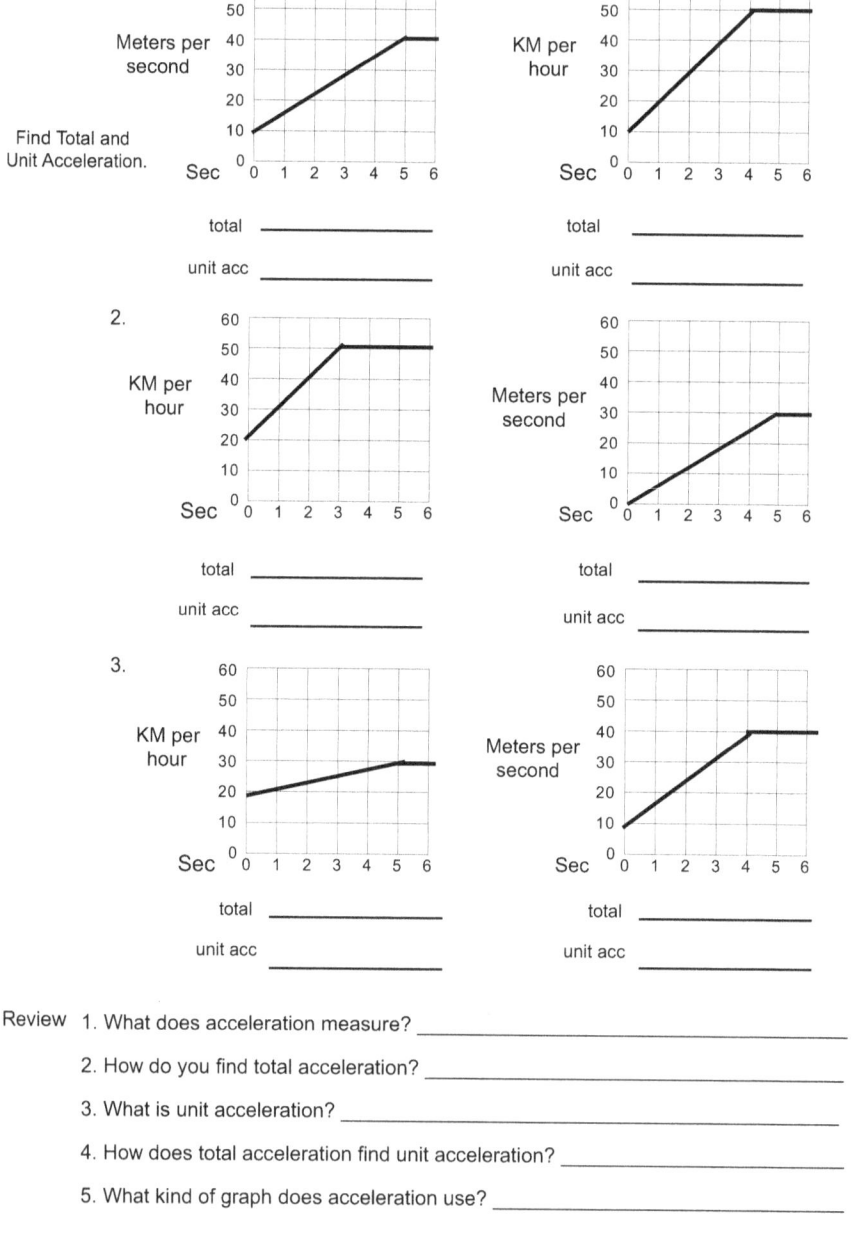

Calculator? yes no

Find Total and Unit Acceleration.

Review
1. What does acceleration measure? _____
2. How do you find total acceleration? _____
3. What is unit acceleration? _____
4. How does total acceleration find unit acceleration? _____
5. What kind of graph does acceleration use? _____

Ch 7 Ls 2   Acceleration Formula.   63

_____ #1 #2 ____/ 7   #3 ____/ 4   R ____/ 6   Total ____/ 17 _____
         Name   Checker

#1 1. What is the acceleration formula? _____
       2. How does acceleration formula change if it doesn't start at 0? _____
       3. How is rate formula different from acceleration? _____
       4. What kind of graph does acceleration use? _____

#2 1. What is happening in this equation? m/sec^2 sec m/sec
                                                $y = 10t + 30$

                  Start _____ then _____ .   Solve t is 5.

                  Solve them. $y = 10$ ___ $+ 30$

                               Rate _____

       2. What is happening in this equation?   kph/s sec kph
                                                $y = 12t + 20$

                  Start _____ then _____ .   Solve t is 6.

                  Solve them. $y = 12$ ___ $+ 20$

                               Rate _____

       3. What's the equation?   Mr W drove 30 kph, then sped up
                                 at 4 meters per second for 8 seconds.

                                           _____   Solve for 8 seconds.

       What does the graph look like?   _____

                                        0 1 2 3 4 5 6 7 8 9 10

                                   _____

#3

1.

M per second — graph rising from 30 at 0 sec to 50 at 4 sec, level to 6 sec

y = (___ x ___) + ___
y = ___ at 5 sec

KM per hour — graph rising from 10 at 0 sec to 40 at 3 sec, level to 6 sec

y = (___ x ___) + ___
y = ___ at 3 sec

Calculator?   yes   no

2.

KM per hour — graph rising from 20 at 0 sec to 30 at 5 sec, level to 6 sec

y = (___ x ___) + ___
y = ___ at 5 sec

M per second — graph rising from 30 at 0 sec to 50 at 5 sec, level to 6 sec

y = (___ x ___) + ___
y = ___ at 4 sec

Review
1. What is the acceleration formula? _____
2. How does acceleration formula change if it doesn't start at 0? _____
3. How is rate formula different from acceleration? _____
4. What kind of graph does acceleration use? _____

Calculator?   yes   no

5. 20 m/sec, 20 m/sec^2 for 6 sec.   30 m/sec, 10 m/sec^2 for 5 sec.

Graph the equation, then find total and unit acceleration.

Km per hour — blank grid, Sec 0–6

total _____
unit acc _____

M per second — blank grid, Sec 0–6

total _____
unit acc _____

Ch 7 Ls 3 Find average speed. 65

_____ #1 #2 ____ / 9   #3 ____ / 4 R ___ / 8 T ___ / 21 _____
         Name   Checker

#1 1. Add start and end speed. Divide by 2. What does it find? _____
    _____

    2. What does it equal? _____
    3. How do you use average average rate? _____
        4. 1st step to average?   **Accelerate from 20 kph to 40 kph in 5 sec**

            2nd step to average? Label it.   ____ + ____ = _____

                                          ____ ÷ ____ = ____ _____

        5. 1st step to average?   **Accelerate from 15 kph to 35 kph in 6 sec**

            2nd step to average? Label it.   ____ + ____ = _____

                                          ____ ÷ ____ = ____ _____

#2 1. What does the 1st equatrion find? _____
    2. What equation does the 1st use to find it? _____
    3. What does the 2nd equation find? _____
        4. Make an equation. Each second a jet uses 150 m of runway to slow down 30
                              kph. It lands at 330 kph. Is a 600 m runway long enough?

                What's the 1st step?   _____

                What's the 2nd step?   _____

        Solve to see if 600 m is long enough.   _____

                                                _____

                                                _____

#3 1. Each second a jet uses 40 meters of runway to slow down 50 kph. It lands at 250 kph. Is a 500 m runway long enough? (You need an equation and solve it for an answer.)

_____
_____
_____
_____
_____

Calculator? yes   no

2. Ojas started his motorcycle and he travels at 30 m/sec, but had to stop at 10 m/s/s to avoid a rabbit. The rabbit was 50 m away. Did he get away? (You need an equation and solve it for an answer.)

_____
_____
_____
_____
_____
_____

Review 1. Add start and end speed. What finds average acceleration? _____

Calculator? yes   no

2. What does it solve to find? _____

3. What does the 1st equation find in the jet problem? _____

4. What does the 2nd equation find? _____

5. How did the 2nd equation use the answer from the 1st? _____

What's Happening?

6.   m/sec m/sec
   $$\frac{10 + 40}{2} = r_{average}$$

_____
_____

7.   m/sec m/sec
   $$\frac{20 + 70}{2} = r_{average}$$

_____
_____

8.   kph/sec   kph
   $$0 = 25x - 275$$
   sec meters total
   $$11 \bullet 40 = t$$

_____
_____

Solve it. Is a 500 meter runway long enough?

Review Problems 67

_____ #1 #2 #3 ____/ 20   #3 #4 ____/ 7   T ____/ 27
            Name

#1 1. Total Acceleration _____
    2. Unit Acceleration _____
    3. Instantaneous Rate _____
    4. Acceleration Equation _____
    5. Acceleration Graph _____
    6. Constant Rate _____

#2. Use Slope and Y intercept to solve each equation.   Calculator?
                                                                 yes   no

1. **Accelerate 10 - 40 kph in 7 sec**   total _____   Unit _____
2. **Accelerate 20 - 50 kph in 10 sec**   total _____   Unit _____
3. **Accelerate 40 - 60 kph in 4 sec**   total _____   Unit _____
4. **Accelerate 10 - 30 kph in 5 sec**   total _____   Unit _____

#3 1.   M per second (graph 0-120, rising to 100 at sec 3, flat)

Find total and accelerations.

total _____

unit acc _____

Kilometers per hour (graph 0-60, rising from 20 to 30 over sec 0-5)   Calculator? yes no

total _____

unit acc _____

2. M per second (graph 0-60, rising to 50 at sec 3, flat)

total _____

unit acc _____

Ft per second (graph 0-120, rising from 60 to 80 at sec 3, flat)

total _____

unit acc _____

#3 Make an equation for the graph and solve it.   Calculator? yes   no

1.

Meters per second

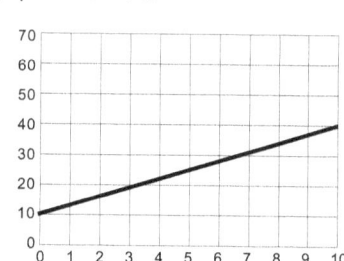

KiPH

y = (___ x ___) + ___

y = ___

y = (___ x ___) + ___

y = ___

2.

Meters per second

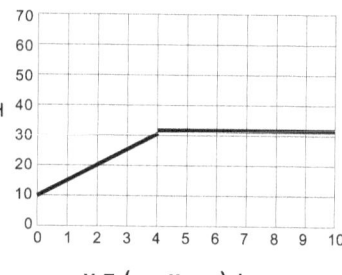

MKPH

y = (___ x ___) + ___

y = ___

y = (___ x ___) + ___

y = ___

#4 1. Each second a jet uses 80 meters of runway to slow down 30 kph. It lands at 300 kph. Is a 2000 m runway long enough?

yes   no

Calculator? yes   no

2. Mitul travels at 60 m/sec, but had to stop at 15 m/s/s for a rabbit. The rabbit was 40 m away. Did the rabbit get away?

3. Mrs J gave a test. The lowest grade was 67 and the highest was 105. What was the average?

Ch 8 Ls 1 How inequalities use a variable. 69

_____ #1 #2 ____/ 9   #3 ____/ 6   R ___/ 12 Total ____/ 27 _____
　　　　Name　　　　　　　　　　　　　　　　　　　　　　　　　　　　　　　　　　　　　Checker

#1 1. How does a variable change an inequality? **x < 5** _____

　　2. When does an inequality use a number line for the graph? _____

　　3. How does it show 1 number and all the numbers? _____

　　4. What does a solid and hollow dot mean at the end of a solid line? _____

　　　　　　5. Solve the equation.　　　　　　**2x - 1 > 3**

　　　　　　　　Graph it.

　　　　　　　　　　　　　　　0 1 2 3 4 5 6 7 8 9 10 11

　　　　　　6. Solve the equation.　　　　　　**- x ≤ - 4**

　　　　　　　　Graph it.

　　　　　　　　　　　　　　　0 1 2 3 4 5 6 7 8 9 10 11

#2 1. Make an equation. What kind
　　　of numbers is it graphing? Does　　0 1 2 3 4 5 6 7 8 9 10 11
　　　it use All Numbers or integers?

　　Circle one.　All Numbers　Integers

　　　2. What's the equation?　　　　　0 1 2 3 4 5 6 7 8 9 10 11
　　　　All Numbers or integers?

　　Circle one.　All Numbers　Integers

　　　3. What's the equation?　　　　　0 1 2 3 4 5 6 7 8 9 10 11
　　　　All Numbers or integers?

　　Circle one.　All Numbers　Integers

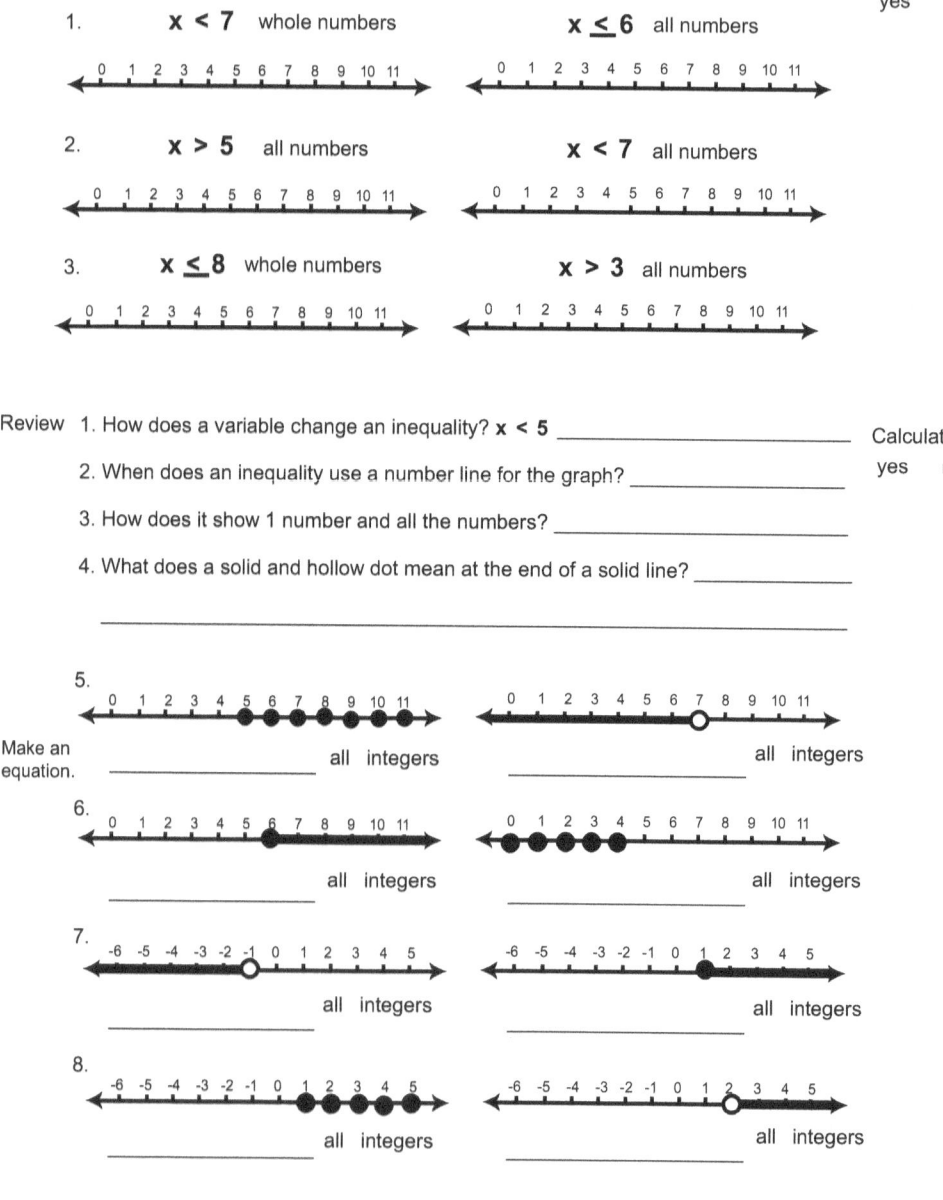

Ch 8 Ls 2 How inequalites use 2 variables. 71

_____   #1 #2 ____ / 8   #3 ____ / 4   R ____ / 7 Total ____ / 19 _____
         Name   Checker

#1 1. What's the 1st step to graph a 2 variable inequality? _____

     2. What is the 2nd step? _____

     3. What's the last step? _____

     4. Name 3 steps for this. _____

     _____

#2 1. What's happening in this equation? $y < 3x + 4$

     Slope: ____ Y intercept: ____ < means _____

     2. What's happening in this equation? $y \geq -4x - 1$

     Slope: ____ Y intercept: ____ $\geq$ means _____

     3. Make an equation for this graph.

     _____

     4. Make an equation for this graph.

     _____

#3 Make a graph for each inequality.   Calculator? yes   no

1.
y < x - 2
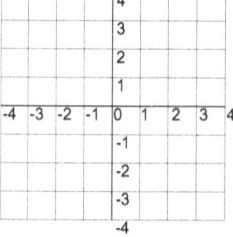

y ≥ 2x - 2
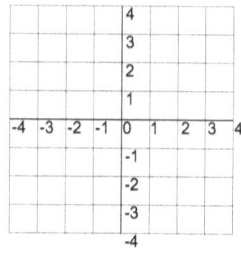

2.
y < x + 1
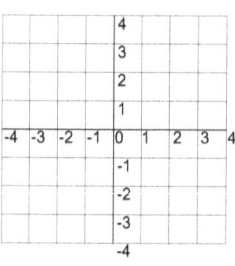

$y > \frac{1}{2} x - 1$
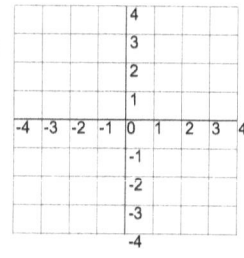

Review
1. How do you solve a 2 variable inequality? _____   Calculator? yes no
2. What's the 1st step to graph a 2 variable inequality? _____
3. What shows what kind of line it is? _____
4. How does the graph show less than or greater than? _____
5. Name 3 steps for this. _____

6. Make an equation.

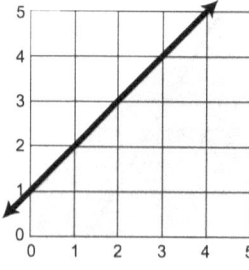

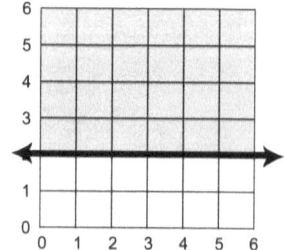

_____   _____

Ch 8 Ls 3 Story problem and inequalities.   73

_____ #1 #2 ____/7   #3 ____/4   R ____/7   T ____/18 _____
    Name   Checker

#1 1. What's the 1st step to make an inequality? _____

2. How do you find another answer? _____

3. How can you tell if it's equal to or not? _____

4. Name 3 steps. _____
_____

#2 1. Find a starting equation.   Anvi has 720 points in science. She needs at least a 1200 for an A. What score does she need on the final test?

    Will it use < or >? _____

    Is it equal to or not? _____

_____

2. Find an equation.   Mr W has more than 500 km to drive. He already went 120 km. He has 5 hours. How fast does he have to go?

    Will it use < or >? _____

    Is it equal to or not? _____

_____

3. Find an equation.   Bongo has 2000 people. They want to have at least 3000 people in 5 yrs. How many people per year is that?

    Will it use < or >? _____

    Is it equal to or not? _____

_____

#3 Inequality story problems.

1. The water tower holds 4,000 kiloliters. It developed a leak at the bottom for 10 liters of water per hour. How long until 1/20 is almost gone? (consider it full)

Calculator? yes no

0 1 2 3 4 5 6 7 8 9 10

2. A car model costs R120,000 and costs an average of Rs 10 to operate per kilometer. A higher model costs Rs 140,000 with an average of Rs 8 to operate. When are they the same?

0 1 2 3 4 5 6 7 8 9 10

Review

1. What's the 1st step to make an inequality? _____
2. How do you find another answer? _____
3. How do you decide which sign to use? _____
4. How can you tell if it's equal to or not? _____
5. Name 3 steps. _____

Calculator? yes no

Solve it.

6.   KPH Hours Kilometers
$60 \cdot t > 150$
How long until Ben is home.

7. KPH Hrs Start Kilometers
$50t + 20 > 270$
How many hours Mr K works.

Review Problems 75

_____ #1 #2 ____ / 10 #3 ____ / 7 Total ____ / 17
Name

#1   1.   **x < 5** whole numbers   **x ≤ 2** all numbers   Calculator?
                                                                              yes   no

         0 1 2 3 4 5 6 7 8 9 10 11   0 1 2 3 4 5 6 7 8 9 10 11

Graph each
inequality. 2. **x > 3** whole numbers   **x < 8** all numbers

         0 1 2 3 4 5 6 7 8 9 10 11   0 1 2 3 4 5 6 7 8 9 10 11

#2 Make an equation for each inequality.

1.   Calculator?
                                                                              yes   no

   y < -x + 3   y > 3x + 1

2.
   y < -x - 1 $y \frac{2}{3}$ x + 3

3.
   y < x + 2 $y \frac{1}{4}$ x - 2

#3 Solve these inequality story problems.

Calculator?
yes   no

1. TJ has bowling scores of 160 and 184. What score does he need in the 3rd set to get more than 520?

2. Eva rented a bike. They charged her Rs 50 per hour, plus a Rs 30 initiaion fee. Ann paid less than Rs 300. Write an inequality for this.

3. Hansh has Rs 5000 to buy an oil change for Rs 4000 and fix a flat for Rs 300 each. How many fix a flat's can he get?

4. B town has 20,000 people and increases at 2,000 per year. U town has 30,000 people and loses 1,000. How long until B town is equal to or greater than U town?

5. An Uber driver charges Rs 50 flat rate plus Rs 30 per km. A customer has R 200 and wants to go 7 km to their home. Can he make it?

6. JJ has Rs 8000 at the beginning of summer. He's allowed to take Rs 500 a week for rent and food. He wants to end with Rs 1000 in his account. How many weeks can he live?

7. If one roll of duct tape costs Rs 275, how many rolls can TJ buy with at least R 800?

Ch 8 Ls 4 How And Problems work. 77

_____ #1 #2 ____/12 #3 ____/6 R ___/ 15 T ____/ 33 _____
Name   Checker

#1   1. How can you tell an **And Problem**? _____
2. What's the first step to rewrite an **And Equation**? _____
3. What does an **And Equation** graph look like? _____
4. How does an **And** Problem show which graph to use? _____
5. Write as 2 inequalities.   - 2 < x < 6

_____   _____   **Describe the graph.**

-10 -8 -6 -4 -2 0 2 4 6 8 10

6. What's the equation for this graph?   -10 -8 -6 -4 -2 0 2 4 6 8 10

_____

#2   1. What signs do **Or Problems** use? _____
"Or"   2. How does the equation show the graph? _____
Equations
3. Name 2 steps to solve both Or and And Problems. _____
4. What does an Or Problem graph look like? _____
5. Write as 2 inequalities.   1 > x > 3

_____   _____   **Describe the graph.**

-10 -8 -6 -4 -2 0 2 4 6 8 10

6. What's the equation for this graph?   -10 -8 -6 -4 -2 0 2 4 6 8 10

_____

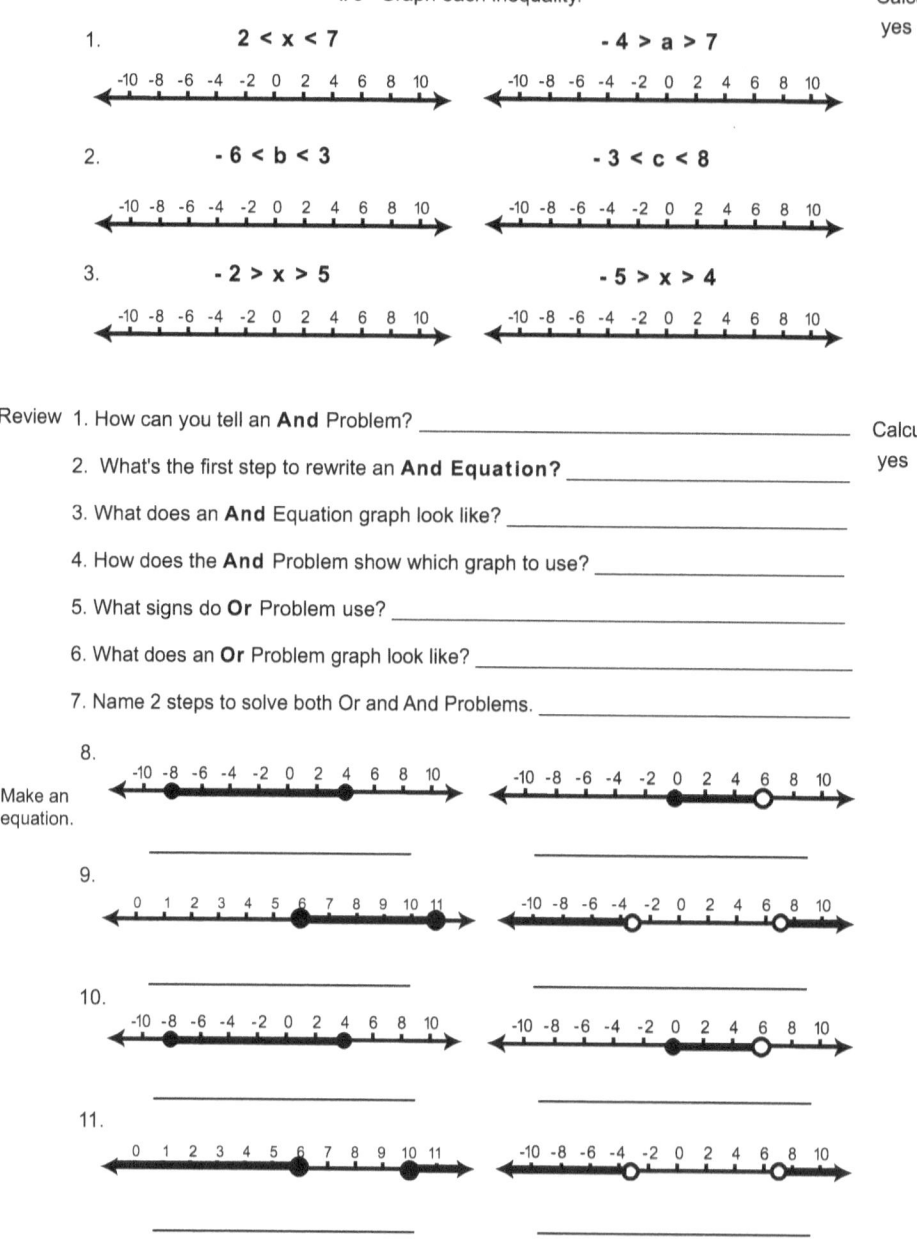

Ch 8 Ls 5 Add a step to compound inequality. 79

_____ #1 #2 ____/ 12 #3 ____/ 4   R ___/ 9 T ___/ 25 _____
           Name   Checker

#1 1. Where does a compound problem add a step? _____
2. Name 2 steps to solve the inequality. _____
3. How did Adding 1 change the graph? _____
4. What is the Opposite Rule for adding a step? _____

5. Write as 2 inequalities.   4 > x + 1 > 8

           _____   _____   Solve the step.
                                            Switch the variable.

           _____   _____   Describe each graph.

              -10 -8 -6 -4 -2 0 2 4 6 8 10

#2 Multiply the Middle Term

1. Where does a compound problem multiply a step? _____
2. What's the 1st step to solve a multiplied step? _____
3. How did multiplying it change the graph? _____
4. How did + 2 and 2x move the graph? _____

5. Write as 2 inequalities.   4 < 2x < 8

           _____   _____   Solve the step.
                                            Switch the variable.

           _____   _____   Describe each graph.

              -10 -8 -6 -4 -2 0 2 4 6 8 10

6. How do you move the graph left 1 space? _____
7. If you want a space twice as wide, what happens to the equation? _____

_____

#3 Graph Each Inequality.   Calculator? yes no

1. $3 > x - 1 > 7$ $-4 > 2a > 6$

   _____ _____   _____ _____

   _____ _____   _____ _____

   -10 -8 -6 -4 -2 0 2 4 6 8 10   -10 -8 -6 -4 -2 0 2 4 6 8 10

2. $2 > x + 2 > 8$ $-4 > 3a > 9$

   _____ _____   _____ _____

   _____ _____   _____ _____

   -10 -8 -6 -4 -2 0 2 4 6 8 10   -10 -8 -6 -4 -2 0 2 4 6 8 10

Review

1. Name 2 steps to solve the inequality. _____

2. How does Adding 1 change the graph? _____

3. What is the Opposite Rule for adding a step? _____

4. How did multiplying it change the graph? _____

5. What's the 1st step to solve a multiplied step? _____

6. How does an equation make a space twice as wide? _____

Calculator? yes no

7. Mr K is driving 60 kph from AW to D City, a distance of more than 300 km. After driving 90 km, Mr K stops for gas. Write and solve an inequality to find how much time he has to reach D City.

   _____
   _____
   _____
   _____

8. TJ and Ira are going golfing this evening. TJ wants to have at least Rs 1200 cash in his wallet. He currently has Rs 400. Write and solve an inequality to find how much he needs to withdraw from the bank.

   _____
   _____
   _____
   _____

9. The OSU Club limits the membership to 750. Currently the club has 620 members. Write and solve an inequality to find how many more members can be recruited.

   _____
   _____
   _____
   _____

Review Problems 81

_____ #1 #2 #3 ____ / 18 #4 #5 ____ / 14 Total ____ / 32
                Name

#1 1. Inequality _____
    2. And Problems _____
    3. Or Problems _____
    4. Add a Step Equation _____

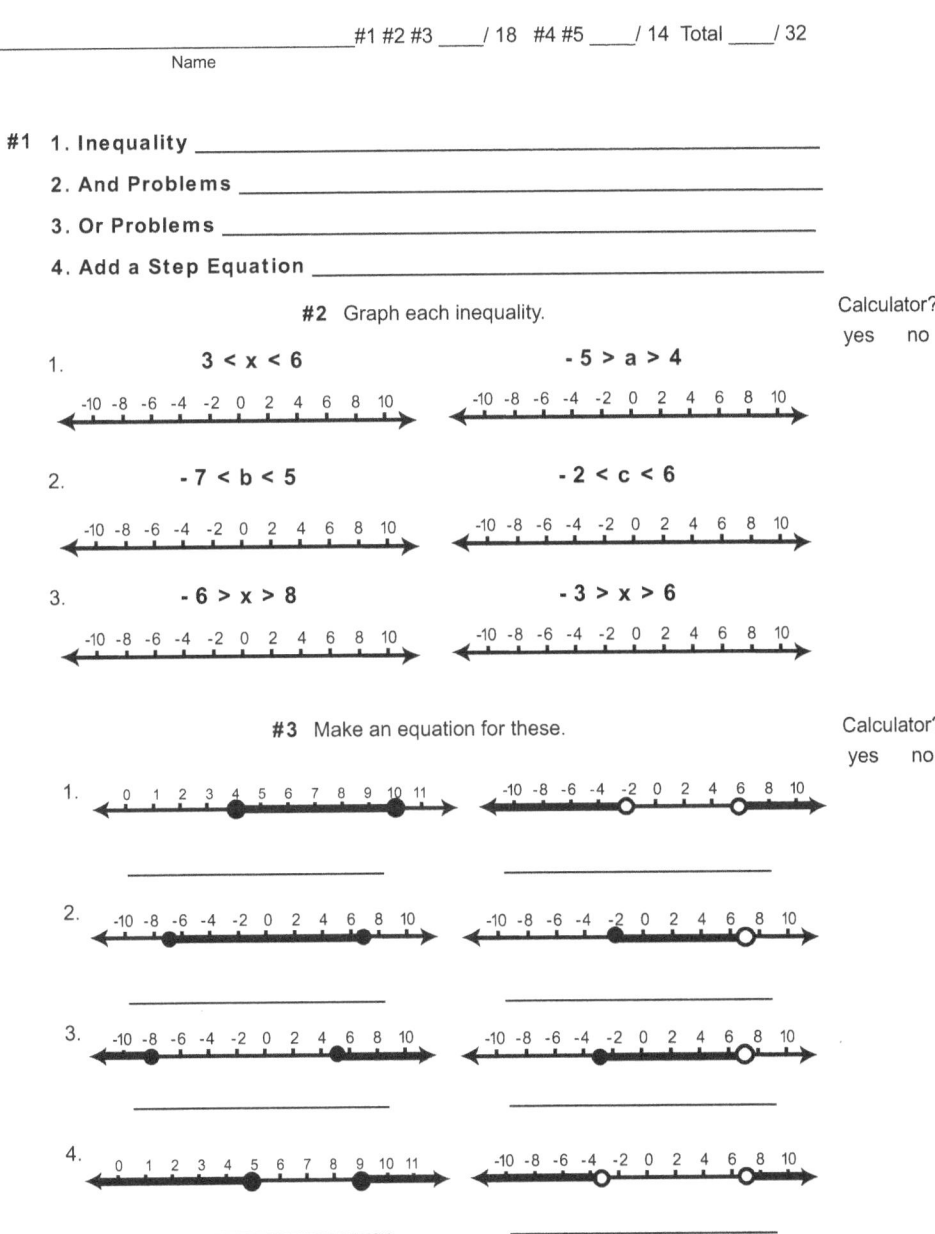

#4 Graph each inequality.   Calculator? yes no

1. $-4 < a < 6$ $-8 > x > 5$

2. $-5 < b < 4$ $-2 < c < 7$

3. $-7 > z > 6$ $-3 > x > 8$

4. $-5 > a - 2 > 6$ $-4 < 3b < 6$

5. $4 > c + 3 > 7$ $-5 > 2x > 8$

#5 Story Problems   Calculator? yes no

1. Eva is baking bread. The recipe calls for more than 3 liters of flour. She has already put in a liter. How many more liters does she need to put in?

2. JJ buys a package of 6 wipers for under Rs 1200. What is the price range for each wiper?

Ch 9 Ls 1 Absolute Value Expressions.   83

_____ #1 #2 ____/12 #3 #4 ____/18 R ___/ 6 Total _____/36 _____
          Name   Checker

#1 1. What is the absolute value of a negative number? _____
     2. How many answers does an absolute value have? _____
     3. What rule solves an absolute value expression? _____
     4. What does absolute value of - 5 equal? | - 5 | _____
          5. Solve the absolute value. | 3 - 5 |
              Negative or Positive?   _____

                                                _____

          6. Solve the 1st step. | - 6 - 3 | + 2 - 5
              Solve it. Negative or Positive?   _____

                                              _____

          7. Solve the 1st step. | 7 - 9 | + 3 - 6
              Solve it. Negative or Positive?   _____

                                              _____

#2 1. How is an absolute value of a variable different? _____
     2. How do you solve this? | x | = 4 _____
     3. What number can't an absolute value equal? _____
         4. What's the 1st step for a variable? | x | - 2 = 5
              Solve it.   _____

                                              _____

          5. What's the 1st step here? | x | + 7 = 6
              Solve it.   _____

                                              _____

#3 What does each expression equal? Calculator? yes no

1. $|-5|$ $|-1|$ $|-8|$

   ___ or ___   ___ or ___   ___ or ___

2. $|2-5|$ $|-4-3|$ $|-2-7|$

   ___ or ___   ___ or ___   ___ or ___

3. $|2-5|+1-5$ $|-1-4|+6-4$ $|7-9|+3-6$

   _____   _____   _____

   ___ or ___   ___ or ___   ___ or ___

#4 1. $|x|=2$ $|x|=4$ $|x|=9$

   X is ___ or ___   X is ___ or ___   X is ___ or ___

2. $|x|-1=5$ $|x|+8=5$ $|x|-2=4$

   _____   _____   _____

   X is ___ or ___   X is ___ or ___   X is ___ or ___

3. $|x|-2=6$ $|x|+2=6$ $|x|+2=4$

   _____   _____   _____

   X is ___ or ___   X is ___ or ___   X is ___ or ___

Review 1. What is the absolute value of a negative number? $|-8|$ _____   Calculator? yes no

2. How does a graph show absolute value answers? _____

3. Do you solve outside or inside an absolute value expression first? _____

4. How is an absolute value of a variable different? _____

5. How do you solve this equation? $|x|=4$ _____

6. What number can't an absolute value equal? _____

Ch 9 Ls 2   Solve absolute value equations.   85

_____ #1 #2 ____/ 8   #3 #4 ____/12 R ____/ 9 Total ____/30 _____
Name   Checker

$$|x + 1| = 3$$

#1   1. What's the 1st step to solve a 1 step absolute value? _____

2. What's the 2nd step to solve it? _____

3. How do you solve of it's multiplied? $|2x| = 4$ _____

4. What's 1st with more steps? $|2x| - 1 = 5$

What happens next?   _____

What you see. What's next?   _____

What are the 2 points?   _____

Describe the graph.   _____

```
  -10  -8  -6  -4  -2   0   2   4   6   8  10
←——|———|———|———|———|———|———|———|———|———|———|——→
```

$$2|x + 1| = 6$$

#2   1. How do you solve an absolute value that's multiplied? _____

2. How do you solve the absolute value? _____

3. How do you solve an absolute value with a fraction? _____

4. What's 1st with a binomial? $3|x + 1| - 1 = 4$

What happens next?   _____

What you see. What's next?   _____

What are the 2 points?   _____

Describe the graph.   _____

```
  -10  -8  -6  -4  -2   0   2   4   6   8  10
←——|———|———|———|———|———|———|———|———|———|———|——→
```

#3 Solve these. What is x?   Calculator? yes no

1. Solve what you see. $|x - 4| = 2$ $|2x| = 6$

   Solve the other way. _____   _____
                        _____   _____
                        _____   _____

   X is ___ or ___   X is ___ or ___

2. Solve what you see. $2|x - 3| = -2$ $4|x - 1| = 3$

   Solve the other way. _____   _____
                        _____   _____
                        _____   _____

   X is ___ or ___   X is ___ or ___

$|x + 1| = 3$

Review
1. What's the 1st step to solve a 1 step absolute value? _____   Calculator?
2. What's the 2nd step to solve it? _____   yes no
3. How do you solve of it's multiplied? $|2x| = 4$ _____
4. How do you solve an absolute value that's multiplied? _____
5. How do you solve the absolute value? _____

6. Solve what you see. $|x - 2| = 3$ $|3x| + 2 = 9$

   Solve the other way. _____   _____
                        _____   _____
                        _____   _____

   Graph it.   _____   _____
                        _____   _____

   -10 -8 -6 -4 -2 0 2 4 6 8 10   -10 -8 -6 -4 -2 0 2 4 6 8 10

Ch 9 Ls 3 2 variable absolute value equations. 87

_____ #1 #2 ____/ 6 #3 #4 ____/ 3 R ___/ 4 Total ____/13 _____
         Name   Checker

#1 1. What does a standard 2 variable absolute value graph look like? _____
    2. How does slope change absolute value graphs? _____
    3. What if slope divides by 2? _____
    4. What if the negative is outside the absolute value? _____

#2 1. Find 3 points. $y = |2x|$ x   0   2   -2

         Describe the graph.   y

    2. Find 3 points. $y = -|4x|$ x   0   1   -1

         Describe the graph.   y

#3 Find 3 points and graph.   Calculator? yes   no

1.
$y = |5x|$

x	-1	0	1
y			

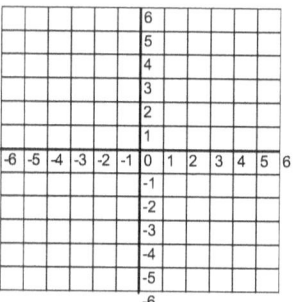

2.
$y = -3|x|$

x	-2	0	2
y			

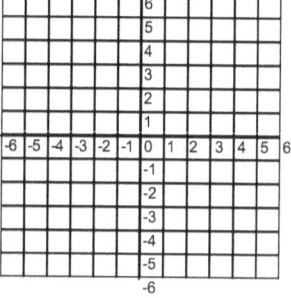

3.
$y = |\tfrac{1}{3}x|$

x	-3	0	3
y			

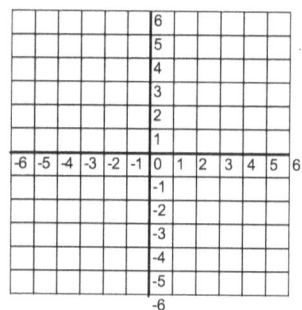

Review 1. What does a standard 2 variable absolute value graph look like? _____
_____

2. How does slope change absolute value graphs? _____
3. What if slope divides by 2? _____
4. What if the negative is outside the absolute value? _____

Ch 9 Ls 3 2 variable absolute value equations. 89

_____ #1 #2 _____ / 19 #3 #4 _____ / 7 Total _____ / 26
Name

#1 What does each expression equal?   Calculator?
                                          yes no

1. $|4-7|$ $|-5-2|$ $|-6-8|$
    ____ or ____   ____ or ____   ____ or ____

2. $|2-8|+7-3$ $|-3-5|+2-6$ $|1-7|+4-9$
    ____ or ____   ____ or ____   ____ or ____

3. $|x|=4$ $|x|=7$ $|x|=15$
    X is ____ or ____ X is ____ or ____ X is ____ or ____

4. $|x|-3=8$ $|x|+9=2$ $|x|-4=3$
    X is ____ or ____ X is ____ or ____ X is ____ or ____

5. $|x|+1=4$ $|x|-4=5$ $|x|-3=5$
    X is ____ or ____ X is ____ or ____ X is ____ or ____

#2 Solve the absolute value equations.   Calculator?
                                             yes no

1. Solve what you see. $|x-4|=3$ Solve what you see. $|2x|=8$
    Solve the other way. _____   Solve the other way. _____
                          _____   _____
    X is ____ or ____   X is ____ or ____

2. Solve what you see. $|x-6|=2$ Solve what you see. $|5x|=20$
    Solve the other way. _____   Solve the other way. _____
                          _____   _____
    X is ____ or ____   X is ____ or ____

#3 Solve these problems.   Calculator?
                                                                  yes   no

1. Solve what $2|x-3|=4$ $|3x|=8$
   you see.

   Solve the _____   _____
   other way. _____   _____

              _____   _____

              X is ____ or ____   X is ____ or ____

2. Solve what $3|x-2|=-4$ $5|x-4|=6$
   you see.

              _____   _____

   Solve the _____   _____
   other way.
              _____   _____

              X is ____ or ____   X is ____ or ____

#4 1. $y=|4x|$

         y

    2. $y=|\frac{1}{2}x|$

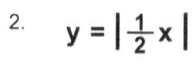

         y

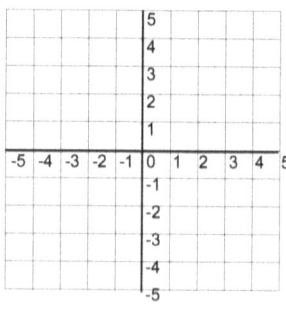

Ch 9 Ls 4 Absolute Value with inequalities.   91

_____ #1 #2 ____/ 6   #3 #4 ____/12   R ___/ 9   Total _____/30 _____
           Name   Checker

#1 1. Name 2 steps to solve absolute value of X. _____

_____

2. What graph is Less than like? _____

3. What graph is Greater than like? _____

#2 1. What's 1st with absolute value x < 5? $|x| < 5$

   What happens next? _____   Solve what you see. X is ____

   What kind of graph is it? _____   Solve the opposite sign. X is ____

                                        -10 -8 -6 -4 -2 0 2 4 6 8 10
   It's between ___ and ___.

2. What's 1st with absolute value x > 4? $|x| > 4$

   What happens next? _____   Solve what you see. X is ____

   What kind of graph is it? _____   Solve the opposite sign. X is ____

   It's greater than ___   -10 -8 -6 -4 -2 0 2 4 6 8 10
   and less than ___.

3. What's 1st with absolute value x + 1 <_ 5? $|x + 1| \leq 5$

   What happens next?   _____

   What are both answers?   _____

   What kind of graph is it? _____

                                        -10 -8 -6 -4 -2 0 2 4 6 8 10
   It's between ___ and ___.

#3 How to solve inequalites. Graph.   Calculator? yes   no

1. What's 1st with this absolute value? $2|x + 1| < 8$

What happens next? _____

What kind of graph is it? _____

   -10 -8 -6 -4 -2 0 2 4 6 8 10

2. What's 1st with absolute value 4x - 1 > 5? $|4x - 1| > 5$

What happens next? _____

What kind of graph is it? _____

   -10 -8 -6 -4 -2 0 2 4 6 8 10

Review

1. Name 2 steps to solve absolute value of X. _____   Calculator? yes   no

2. What graph is Less than like? _____

3. What graph is Greater than like? _____

4. What's 1st with absolute value 2x + 1 < 5? $|2x + 1| < 5$

What happens next? _____

What kind of graph is it? _____

   -10 -8 -6 -4 -2 0 2 4 6 8 10

Ch 9 Ls 5 Y intercept inside absolute variables. 93

_____ #1 #2 ___/ 8   #3 #4 ___/ 2   R ___/ 7   Total ___/ 17 _____
         Name   Checker

#1 1. What's the 1st step to graph an intercept inside? _____

2. What's the 2nd step? _____

3. How does X intercept show how the graph moves? _____

4. What made a graph go negative? _____

5. How do you find the Y intercept? _____

6. How do you find the X intercept? _____

#2 1. Find 3 points. $y = |x - 1|$ x   0   3   -3

Describe the graph.   y

2. Find 3 points. $y = |x + 2| - 3$ x   0   2   -2

Describe the graph.   y

#3 1.

$y = |x| - 2$

Find 3 points and graph.

x	0	2	-2
y			

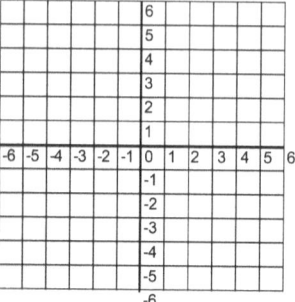

Calculator? yes   no

2.

$y = -|x| + 4$

x	0	4	-4
y			

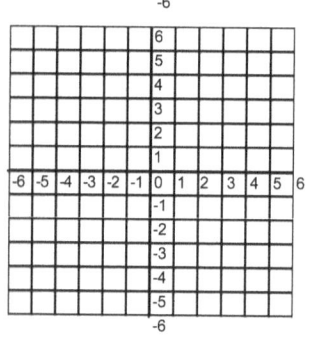

Review 1. What's the 1st step to graph an intercept inside? _____

Calculator? yes   no

2. What's the 2nd step? _____

3. How does X intercept show how the graph moves? _____

4. What made a graph go negative? _____

5. How do you find the Y intercept? _____

6. How do you find the X intercept? _____

7.

$y = |x - 2| - 4$

x	0	2	-2
y			

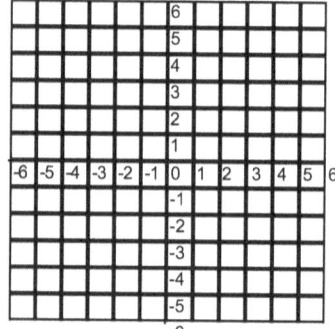

Ch 9 Ls 6 Absolute variable story problems.   95

_____ #1 #2 ____ / 7 #3 ____ / 3 R ___ / 7 Total _____ / 17 _____
　　　　　　Name　　　　　　　　　　　　　　　　　　　　　　　　　　　　　　　　　　　　　Checker

#1 1. What formula finds absolute value equations? _____
　　　　2. What does **m** stand for in the formula? _____
　　　　3. What does **A** stand for? _____
　　　　4. Is **m** always positive or sometimes negative? _____

#2 1. What goes on the right　　　The expected temperature is 30,
　　　　　side of the equation?　　　　plus or minus 2 degrees.

　　　　　What is the left side?

　　　　　Solve the equation.

　　　　2. What goes on the right　　　A wrestler must be + or - 5 kilograms
　　　　　side of the equation?　　　　within his 80 kg weight class.

　　　　　What is the left side?

　　　　　Solve the equation.

　　　　3. What goes on the right　　　A poll says a candidate is
　　　　　side of the equation?　　　　+ or - 2 points of 74% approval.

　　　　　What is the left side?

　　　　　Solve the equation.

#3 Make an equation and solve them.   Calculator?
                                                                        yes   no

1. A wrestler must fall 3 kilograms
   above or below 75 kg.

   Equation _____

   Solve it. _____

   Answer _____

2. A machine shop has 85 millimeter
   part with error rate 0.2 mm.

   Equation _____

   Solve it. _____

   Answer _____

3. A mayor has 58% approval with
   + or - 3% error rate.

   Equation _____

   Solve it. _____

   Answer _____

Review 1. What formula finds absolute value equations? _____   Calculator?
2. What does **m** stand for in the formula? _____   yes   no
3. What does **A** stand for? _____
4. Is **m** always positive or sometimes negative? _____

Solve it.

5.   Middle   degrees
     $|t - 60| > 3$ _____
     Temperature   _____

6.   Middle   Favortism
     $|t - 85| > 2$ _____
     Political Speaker   _____

7.   Middle millimeters
     $|t - 72| > 0.2$ _____
     Part for Car   _____

Review Problems 97

_____ #1 #2 ____ / 8 #3 #4 ____ / 5 Total _____ /13
         Name

#1 1. Absolute Value _____
    2. Absolute Value Expression _____
    3. Absolute Value Equation _____
    4. Absolute Value Inequality _____
    5. Absolute Value Story Problem _____

#2 How to solve inequalites. Graph.   Calculator?
                                                    yes   no
1. What's 1st with this absolute value? $3|x + 2| < 5$

                                      _____
   What happens next?   _____

                                      _____
   What kind of graph is it?   _____
                        -10 -8 -6 -4 -2 0 2 4 6 8 10

2. What's 1st with absolute value 2x - 1 > 6? $|2x - 1| > 6$

                                      _____
   What happens next?   _____

                                      _____
   What kind of graph is it?   _____
                        -10 -8 -6 -4 -2 0 2 4 6 8 10

3. What's 1st with this absolute value? $3|x + 2| < 7$

                                      _____
   What happens next?   _____

                                      _____
   What kind of graph is it?   _____
                        -10 -8 -6 -4 -2 0 2 4 6 8 10

#3 What's 1st with this absolute value?   Calculator? yes no

1. $y = |x - 3| - 1$

x	-3	0	3
y			

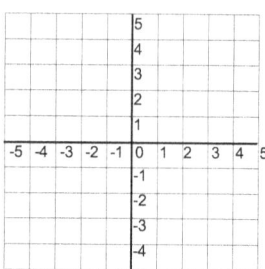

2. $y = |x + 2| - 3$

x	-2	0	2
y			

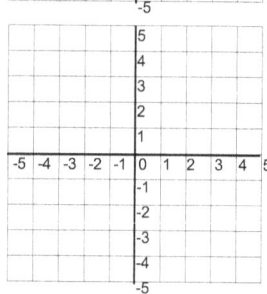

3. $y = -|x - 4| - 2$

x	-4	0	4
y			

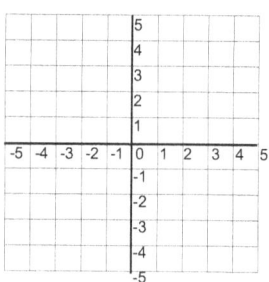

#4 1. Your aquarium's thermometer is accurate to + or - 0.5 degrees C. You want your aquarium to be 23 degrees. What's the lowest and highest temperature for it?   Calculator? yes no

2. Cocoa powder is roasted cocoa beans. The ideal temperature for roasting is 207 degrees C, plus or minus 3 degrees. What is the coolest and warmest the oven should be?

Ch 10 Ls 1   Begin Functions.   99

_____ #1 #2 ____/12 #3 ____/ 4 R ____/ 10 Total _____/ 26 _____
          Name   Checker

#1   1. What is a function? _____

2. How are discrete functions different from continuous ones? _____

_____

3. What test checks if it is a function? _____

4. Are these points a function?   1, 0   2, 5   3, 4

Yes, it is a function.   No, it is not a function.

5. Is this a function?   1, 3   2, 4   3, 3

Yes, it is a function.   No, it is not a function.

#2   1. How do you write a function? _____

2. How is an equation written to show it's a function? _____

3. What is a specific function? _____

4. How do you solve a specific function? _____

5. What's the answer?   a(2) if a(x) = 3x + 1

Finish the solution.   _____

                                _____

6. What's the answer?   b(-2) if b(x) = 2x - 9

Finish it.   _____

                                _____

7. What's the answer?   c(4) if c(x) = $\frac{1}{2}$x + 3

Finish it.   _____

                                _____

#3 Solve these specific functions.

Calculator? yes no

1. a(4) if a(x) = 3x - 2   b(-2) if b(x) = 5x - 8

   _____   _____

   _____   _____

2. c(4) if c(x) = $\frac{1}{2}$x + 3   d(-5) if d(x) = 4x - 9

   _____   _____

   _____   _____

Review 1. What is a function? _____

Calculator? yes no

2. How are discrete functions different from continuous ones? _____

_____

3. What test checks if it is a function? _____

4. How do you write a function? _____

5. How is an equation written to show it's a function? _____

6. What is a specific function? _____

7. How do you solve a specific function? _____

If it's a function, write the equation for it.

8.
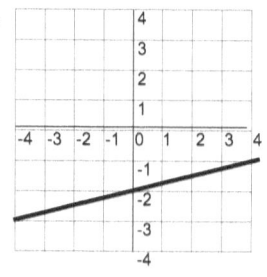 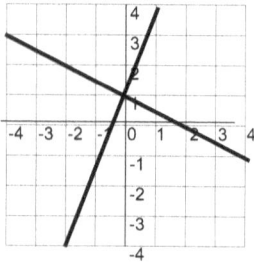 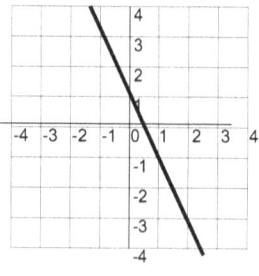

a(x) = _____   b(x) = _____   c(x) = _____

Ch 10 Ls 2 How functions use domain and range.   101

_____ #1 #2 ____/ 7 #3 ____/ 6 R ____/ 12 Total _____/ 25 _____
         Name   Checker

#1 1. What is the domain of a function? _____

2. What is the range of a function? _____

3. How does a graph show if a point in included or not? _____

4. What do arrows show about a limit? _____

#2 1. What is the domain? Find the range.

Domain: _____   Range: _____

2. Find the range. What is the domain?

Domain: _____   Range: _____

3. What is the domain? Find the range.

Domain: _____   Range: _____

#3

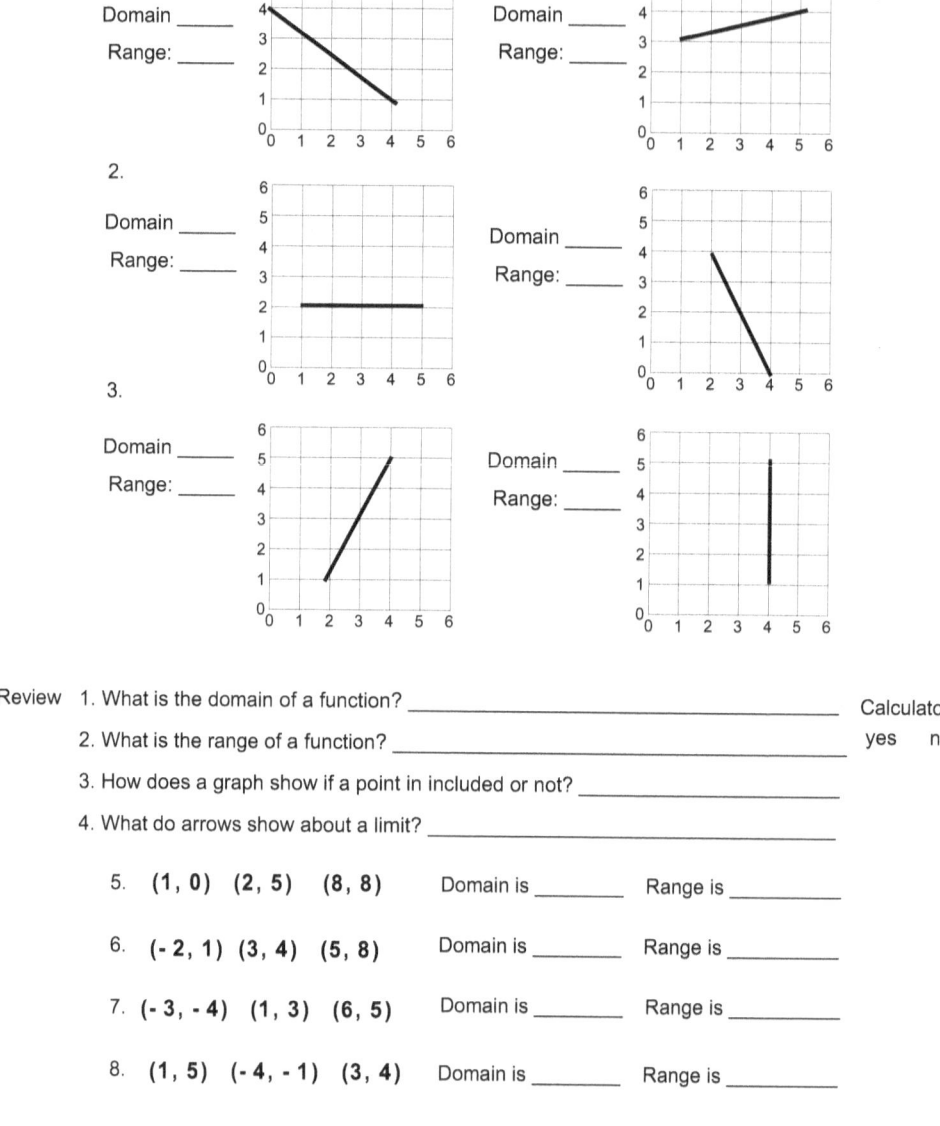

1.
Domain _____
Range: _____

Domain _____
Range: _____

Calculator?
yes   no

2.
Domain _____
Range: _____

Domain _____
Range: _____

3.
Domain _____
Range: _____

Domain _____
Range: _____

Review 1. What is the domain of a function? _____
2. What is the range of a function? _____
3. How does a graph show if a point in included or not? _____
4. What do arrows show about a limit? _____

Calculator?
yes   no

5. (1, 0) (2, 5) (8, 8)   Domain is _____   Range is _____

6. (- 2, 1) (3, 4) (5, 8)   Domain is _____   Range is _____

7. (- 3, - 4) (1, 3) (6, 5) Domain is _____   Range is _____

8. (1, 5) (- 4, - 1) (3, 4) Domain is _____   Range is _____

Ch 10 Ls 3 How functions use 4 operations.   103

_____   #1 #2 ____/ 16 #3 ____/ 8 R ___/ 15 T___/ 39 _____
Name   Checker

#1 1. How does a function show addition and multiplation? _____
    2. For which operations is order important? _____
    3. Why are there 2 ways to subtract functions? _____
    4. What do you use to divide functions? _____

#2 1. Add (a + b)(x) $a(x) = 2x + 2$ $a(x) = -4x + 2$
                            $b(x) = 5x - 5$ $b(x) = 7x - 1$

            _____   _____

    2. Subtract (a - b)(x) $a(x) = -3x + 1$ $a(x) = 2x + 2$
                              $b(x) = 4x - 5$ $b(x) = 6x - 1$

            _____   _____

    3. Multiply (c•d)(x) $c(x) = 7x + 2$ $c(x) = 6x + 3$
                              $d(x) = 4$ $d(x) = 2x$

            _____   _____

    4. Divide $(\frac{c}{d})(x)$ $c(x) = 8x + 2$ $c(x) = 6x + 2$
                                    $d(x) = 4$ $d(x) = 3$

            _____   _____

    5. Add (a + b)(x) $a(x) = 2x + 2$ $a(x) = -4x + 2$
                               $b(x) = 5x$ $b(x) = 8x$

            _____   _____

    6. Subtract (a - b)(x) $a(x) = x - 5$ $a(x) = 4x - 2$
                               $b(x) = 2x - 2$ $b(x) = 7x - 3$

            _____   _____

#3   Add, subtract, and multiply these functions.   Calculator?
                                                            yes   no

1. a(x) = -5x + 4   a(x) = 8x - 2
    + b(x) = -2x - 3 _____   + b(x) = 3x - 4 _____

2. c(x) = 7x + 2   c(x) = -4x - 1
    - d(x) = 5x - 1 _____   - d(x) = -x + 3 _____

3. e(x) = 2x - 5   e(x) = 4x + 2
    • f(x) = 3x _____   • f(x) = -3x _____

4. Divide $\dfrac{c(x) = 15x - 3}{d(x) = 5}$ _____ $\dfrac{c(x) = 8x + 1}{d(x) = 4}$ _____

Review 1. How does a function show addition and multiplation? _____   Calculator?
2. For which operations is order important? _____   yes   no
3. Why are there 2 ways to subtract functions? _____
4. What do you use to divide functions? _____

Divide the   5. a(x) = 2x + 1   e(x) = 2x - 4
functions.   b(x) = 6x _____   f(x) = 0.5 _____

                 6. Which operation   c(x) = 7x + 2
                    did it use?   d(x) = 5x + 1   = 12x + 3   _____
Decide which
operation it used.
                 7. Which operation?   c(x) = 2x + 1
                                        d(x) = 5   = 10x + 5   _____

                 8. Which operation?   c(x) = 7x + 2
                                        d(x) = 5x   = 2x + 2   _____

                 9. Which operation?   c(x) = 2x + 1
                                        d(x) = 4   = 8x + 4   _____

Ch 10 Ls 4 Transfomations   105

_____ #1 #2 ____/ 7 #3 ____/ 12 R ___/ 4 Total ____/ 23 _____
Name   Checker

#1 1. Name 3 ways to transform a line. _____
    2. How does a line shift? _____
    3. Name 2 ways to reflect a line. _____
    4. What is correct word for a point that is dizzy? _____

#2 1. Name 2 ways the functions change.

    x axis _____   y axis _____

    2. How did this function change?

    x axis _____   y axis _____

    3. How did this function change?

    x axis _____   y axis _____

#3 How did the shape move?
(It moves towards the solid line.)

Calculator?   yes   no

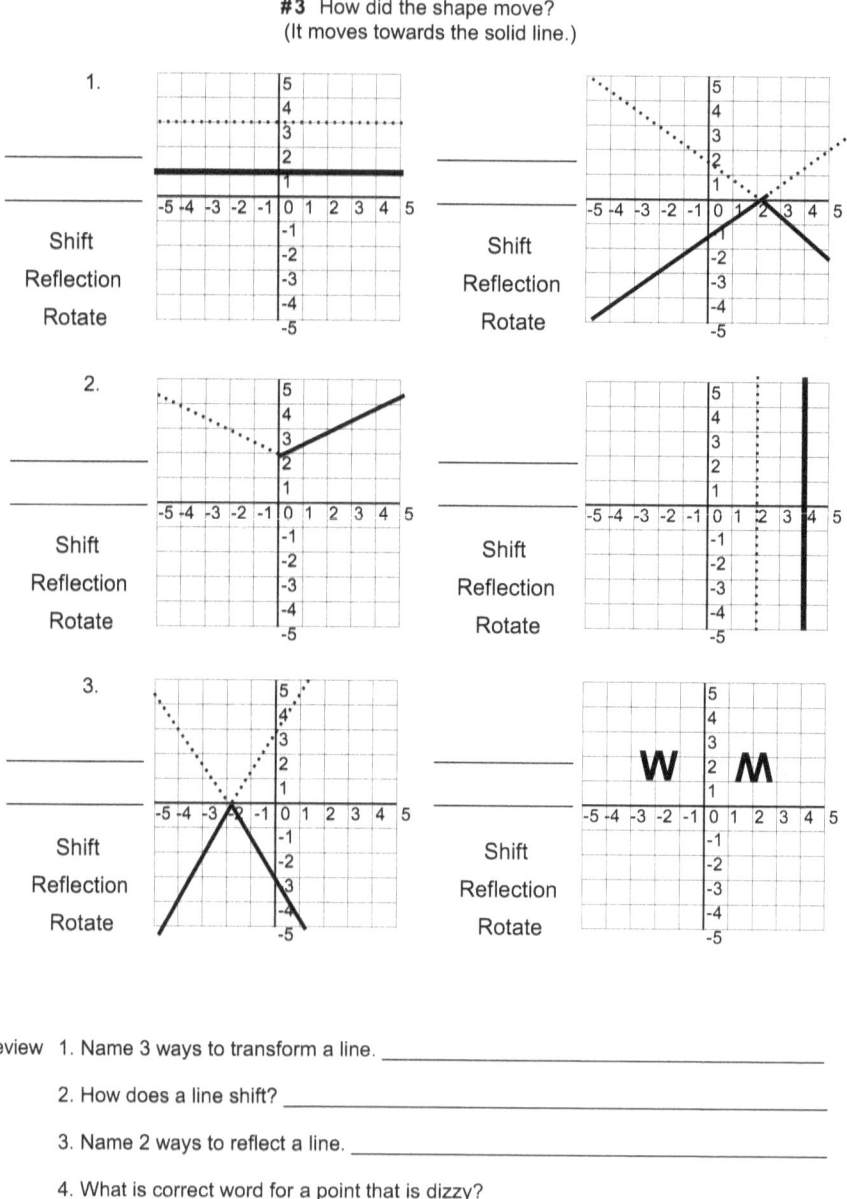

1. Shift / Reflection / Rotate

    Shift / Reflection / Rotate

2. Shift / Reflection / Rotate

    Shift / Reflection / Rotate

3. Shift / Reflection / Rotate

    Shift / Reflection / Rotate

Review 1. Name 3 ways to transform a line. _____

2. How does a line shift? _____

3. Name 2 ways to reflect a line. _____

4. What is correct word for a point that is dizzy? _____

Review Problems   107

_____   #1 #2 #3 ____ / 21 #4 #5 ____ / 20 Total ____ / 41
          Name

#1 1. Functions _____
    2. Specific Functions _____
    3. Domain and Range _____
    4. Shift Transformation _____
    5. Reflective Transformation _____

#2 1. a(6) if a(x) = 4x − 5   b(7) if b(x) = 5x − 12   Calculator?
Solve these   yes   no
problems   _____   _____

                _____   _____

    2. c(5) if c(x) = $\frac{3}{4}$x + 2   d(−3) if d(x) = 9x − 3

                _____   _____

                _____   _____

#3 1.   Calculator?
                                                                          yes   no
    Domain _____   Domain _____
    Range: _____   Range: _____
Find the domain
and range.

    2.
    Domain _____   Domain _____
    Range: _____   Range: _____

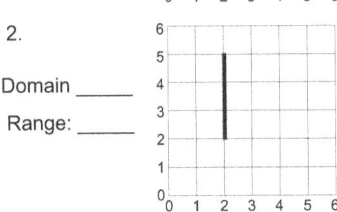

    3.
    Domain _____   Domain _____
    Range: _____   Range: _____

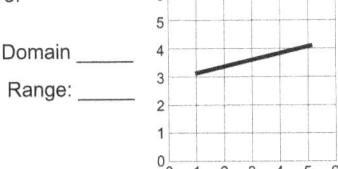

#4 Add, subtract, multiply, and divide these functions.   Calculator? yes no

1. a(x) = -4x + 9   a(x) = 6x - 3
  + b(x) = -3x - 3 _____ + b(x) = 7x - 4 _____

2. c(x) = 8x + 2   c(x) = -5x - 10
  - d(x) = 6x - 7 _____   - d(x) = -7x + 3 _____

3. e(x) = 3x - 6   e(x) = 5x + 3
  • f(x) = 4x _____   • f(x) = -2x _____

4. Divide c(x) = 20x - 5 / d(x) = 2 _____   c(x) = 12x - 1 / d(x) = 15 _____

#5 How did the shape move? (It moves towards the solid shape.)   Calculator? yes no

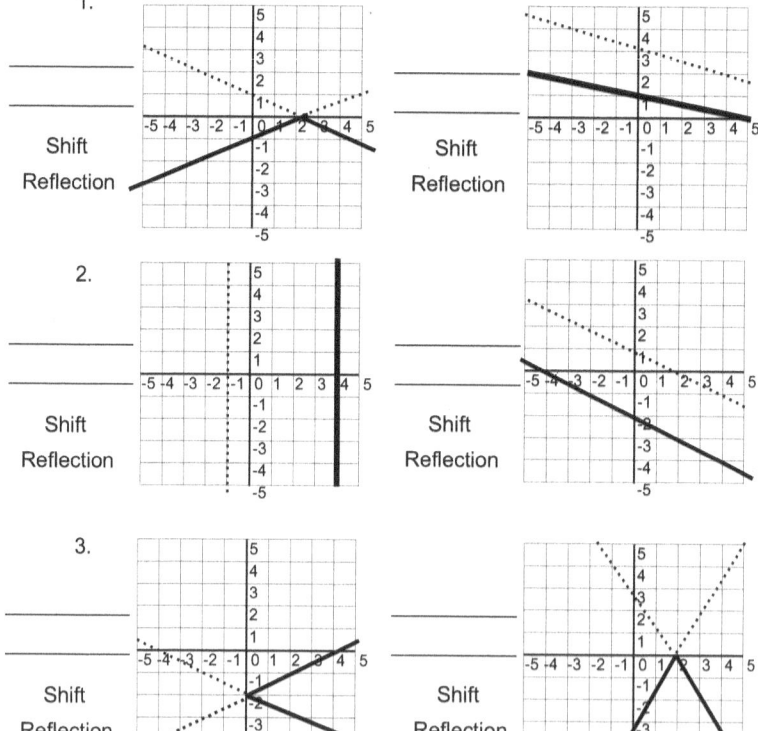

1. _____ Shift _____ Reflection _____   _____ Shift _____ Reflection _____

2. _____ Shift _____ Reflection _____   _____ Shift _____ Reflection _____

3. _____ Shift _____ Reflection _____   _____ Shift _____ Reflection _____

Ch 11 Ls 1 What are piecewise functions?   109

_____ #1 #2 ____ / 5   #3 ____ / 8   R ____ / 5   T ____ / 18 _____
            Name   Checker

#1 1. What are piecewise functions? _____

2. What are 2 parts to piecewise functions? _____

3. What do Piecewise Functions use for domains? _____

#2 1. What's the left side equation and domain?

$$f(x) = \begin{cases} \underline{\hspace{2cm}} \text{ if } \underline{\hspace{2cm}} \\ \underline{\hspace{2cm}} \text{ if } \underline{\hspace{2cm}} \end{cases}$$

Right side?

What's the key point?

Key point is _____.

2. What's the left side equation and domain?

$$f(x) = \begin{cases} \underline{\hspace{2cm}} \text{ if } \underline{\hspace{2cm}} \\ \underline{\hspace{2cm}} \text{ if } \underline{\hspace{2cm}} \end{cases}$$

Right side?

What's the key point?

Key point is _____.

#3 Make the Equations and Domains.   Calculator? yes   no

1.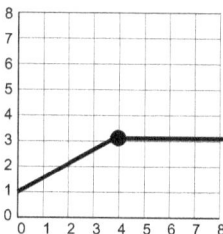

$f(x) = \begin{cases} \rule{2cm}{0.4pt} \text{ if } \rule{1.5cm}{0.4pt} \\ \rule{2cm}{0.4pt} \text{ if } \rule{1.5cm}{0.4pt} \end{cases}$ $f(x) = \begin{cases} \rule{2cm}{0.4pt} \text{ if } \rule{1.5cm}{0.4pt} \\ \rule{2cm}{0.4pt} \text{ if } \rule{1.5cm}{0.4pt} \end{cases}$

2.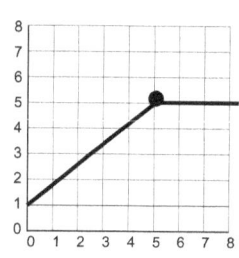

$f(x) = \begin{cases} \rule{2cm}{0.4pt} \text{ if } \rule{1.5cm}{0.4pt} \\ \rule{2cm}{0.4pt} \text{ if } \rule{1.5cm}{0.4pt} \end{cases}$ $f(x) = \begin{cases} \rule{2cm}{0.4pt} \text{ if } \rule{1.5cm}{0.4pt} \\ \rule{2cm}{0.4pt} \text{ if } \rule{1.5cm}{0.4pt} \end{cases}$

Review 1. What are piecewise functions? _____   Calculator? yes   no

2. What are 2 parts to piecewise functions? _____

3. What do Piecewise Functions use for domains? _____

4. $f(x) = \begin{cases} x + 1 & \text{if } x < 2 \\ 3 & \text{if } x \geq 2 \end{cases}$ _____

5. $f(x) = \begin{cases} 2x + 3 & \text{if } x < 2 \\ 7 & \text{if } x \geq 2 \end{cases}$ _____

Ch 11 Ls 2 How functions use And inequalities.   111

_____ #1 #2 ____ / 7   #3 ____ / 4   R ____ / 6   T ____ / 17 _____
               Name   Checker

#1 1. What do Piecewise functions use to stop a function? _____

2. What do you use to find a Y intercept? _____

3. What are 3 parts to the rule? _____

4. What equation do you use to find it? _____

#2 1. What are the left and right side equation and domain?

What's the key point?

Key point is _____ $f(x) = \begin{cases} \underline{\hspace{2cm}} & \text{if } \underline{\hspace{1cm}} \\ \underline{\hspace{2cm}} & \text{if } \underline{\hspace{1cm}} \end{cases}$

2. What are the left and right side equation and domain?

What's the key point?

Key point is _____ $f(x) = \begin{cases} \underline{\hspace{2cm}} & \text{if } \underline{\hspace{1cm}} \\ \underline{\hspace{2cm}} & \text{if } \underline{\hspace{1cm}} \end{cases}$

3. What are the left and right side equation and domain?

What's the X for the key point?

Key point is _____ $f(x) = \begin{cases} \underline{\hspace{2cm}} & \text{if } \underline{\hspace{1cm}} \\ \underline{\hspace{2cm}} & \text{if } \underline{\hspace{1cm}} \end{cases}$

#3 Make the Equations and Domains.   Calculator? yes   no

1.

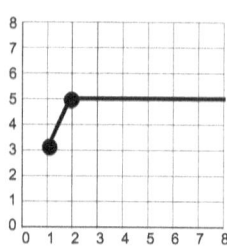

$f(x) = \begin{cases} \underline{\qquad} & \text{if } \underline{\qquad} \\ \underline{\qquad} & \text{if } \underline{\qquad} \end{cases}$

$f(x) = \begin{cases} \underline{\qquad} & \text{if } \underline{\qquad} \\ \underline{\qquad} & \text{if } \underline{\qquad} \end{cases}$

2.

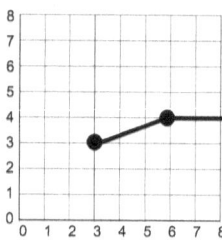

$f(x) = \begin{cases} \underline{\qquad} & \text{if } \underline{\qquad} \\ \underline{\qquad} & \text{if } \underline{\qquad} \end{cases}$

$f(x) = \begin{cases} \underline{\qquad} & \text{if } \underline{\qquad} \\ \underline{\qquad} & \text{if } \underline{\qquad} \end{cases}$

Review 1. What do Piecewise functions use to stop a function? _____   Calculator? yes   no

2. What do you use to find a Y intercept? _____

3. What are 3 parts to the rule? _____

4. What equation do you use to find it? _____

What's Happening?

5. $f(x) = \begin{cases} 2x + 2 & \text{if } 1 \le x < 2 \\ 6 & \text{if } x \ge 2 \end{cases}$ _____

6. $f(x) = \begin{cases} x + 6 & \text{if } 2 \le x < 4 \\ 2 & \text{if } x \ge 4 \end{cases}$ _____

Ch 11 Ls 3 What happens when 2 lines don't meet.   113

_____ #1 #2 ____ /7 #3 ____ /4 R ____ /6 T ____ /17 _____
        Name   Checker

#1 1. What is it called when lines don't meet each other? _____

      2. How does a discontinuity stay a function? _____

      3. Does X or Y show the key point? _____

#2 1. Solve these functions. $f(x) = \begin{cases} x + 1 & \text{if } x < 2 \\ 3 & \text{if } 2 \leq x < 6 \end{cases}$
        Is there discontuity?

_____

2. Solve these functions. $f(x) = \begin{cases} 2x + 3 & \text{if } x < 2 \\ 3 & \text{if } 2 \leq x < 6 \end{cases}$
   Is there discontuity?

_____

3. What's the left side equation?
   The right side equation?

What's the X for the key point?

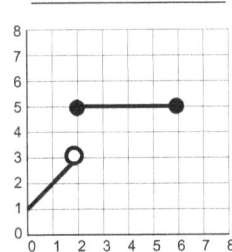

$f(x) = \begin{cases} \underline{\hspace{2cm}} & \text{if } \underline{\hspace{1.5cm}} \\ \underline{\hspace{2cm}} & \text{if } \underline{\hspace{1.5cm}} \end{cases}$

4. What's the left side equation?
   The right side equation?

What's the X for the key point?

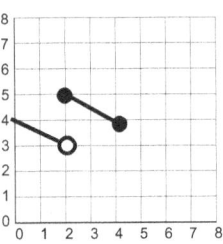

$f(x) = \begin{cases} \underline{\hspace{2cm}} & \text{if } \underline{\hspace{1.5cm}} \\ \underline{\hspace{2cm}} & \text{if } \underline{\hspace{1.5cm}} \end{cases}$

#3 Make the Equations and Domains.   Calculator? yes   no

1.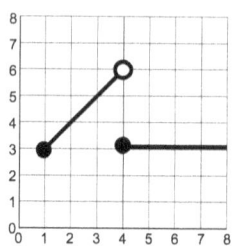

$f(x) = \begin{cases} \underline{\hspace{1cm}} & \text{if } \underline{\hspace{1cm}} \\ \underline{\hspace{1cm}} & \text{if } \underline{\hspace{1cm}} \end{cases}$

$f(x) = \begin{cases} \underline{\hspace{1cm}} & \text{if } \underline{\hspace{1cm}} \\ \underline{\hspace{1cm}} & \text{if } \underline{\hspace{1cm}} \end{cases}$

2.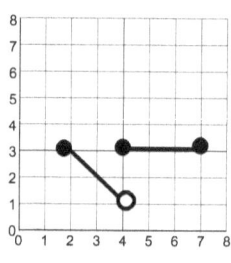

$f(x) = \begin{cases} \underline{\hspace{1cm}} & \text{if } \underline{\hspace{1cm}} \\ \underline{\hspace{1cm}} & \text{if } \underline{\hspace{1cm}} \end{cases}$

$f(x) = \begin{cases} \underline{\hspace{1cm}} & \text{if } \underline{\hspace{1cm}} \\ \underline{\hspace{1cm}} & \text{if } \underline{\hspace{1cm}} \end{cases}$

Review   1. What is it called when the lines don't meet each other? _____   Calculator? yes no

2. How does a discontinuity stay a function? _____

3. Does X or Y show the key point? _____

4. 4 parts to watch for. _____

5. Is there discontinuity?
$f(x) = \begin{cases} 2x + 1 & \text{if } x < 4 \\ 3 & \text{if } 4 \geq x > 6 \end{cases}$ _____

6. $f(x) = \begin{cases} 0.5x + 5 & \text{if } x < 3 \\ 2 & \text{if } x \geq 3 \end{cases}$ _____

Ch 11 Ls 3 Piecewise story problems.   115

_____ #1 #2 ____/ 6   #3 ____/ 4 R ____/ 4 T ____/ 14 _____
                Name   Checker

#1 1. Why are these called "Same Equations"? _____
    2. Why are these called "Different Equations"? _____
    3. How can you tell which kind it is? _____

#2 1. Describe what's happening. (People walk an average of 4 mph.) What would enable Ojas to go that fast?

    kph

    Minutes

2. Describe what's happening to Reya's car.

    m/sec

    seconds

3. Decide how Amav's gas bill got paid.

    Rupees

    months

#3 What's Happening?   Calculator? yes   no

1.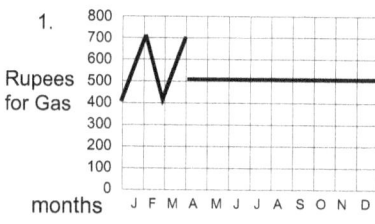

_____

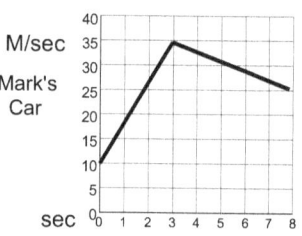

_____

2.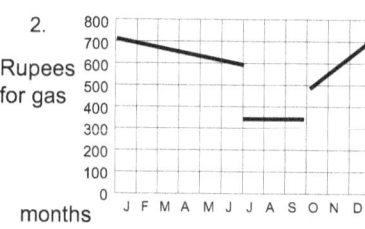

_____

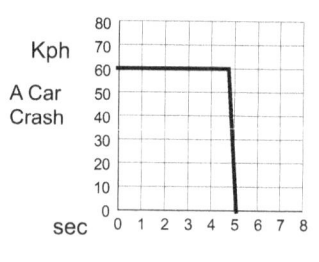

_____

Review 1. How does a graph show you what kind of problem it is? _____   Calculator?

2. What 2 things shows what the line is doing? _____   yes   no

3. How does a graph show constant acceleration? _____

4. What's happening getting on the highway?

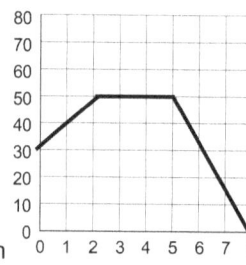

_____
_____

Review Problems 117

_____ #1 #2 ____ / 9   #3 ____ / 9   Total ____ / 18
Name

#1 1. Piecewise Functions _____
    2. Domain _____
    3. Discontinuity _____

#2. 1.

Make the Equations
and Domains

Calculator?
yes   no

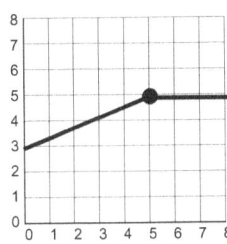 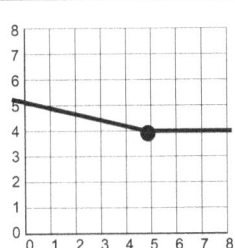

$f(x) = \begin{cases} \underline{\hspace{1cm}} \text{ if } \underline{\hspace{1cm}} \\ \underline{\hspace{1cm}} \text{ if } \underline{\hspace{1cm}} \end{cases}$ $f(x) = \begin{cases} \underline{\hspace{1cm}} \text{ if } \underline{\hspace{1cm}} \\ \underline{\hspace{1cm}} \text{ if } \underline{\hspace{1cm}} \end{cases}$

2.

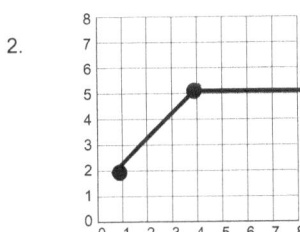 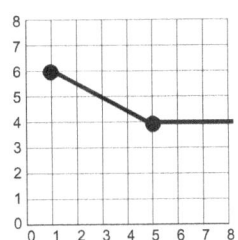

$f(x) = \begin{cases} \underline{\hspace{1cm}} \text{ if } \underline{\hspace{1cm}} \\ \underline{\hspace{1cm}} \text{ if } \underline{\hspace{1cm}} \end{cases}$ $f(x) = \begin{cases} \underline{\hspace{1cm}} \text{ if } \underline{\hspace{1cm}} \\ \underline{\hspace{1cm}} \text{ if } \underline{\hspace{1cm}} \end{cases}$

3.

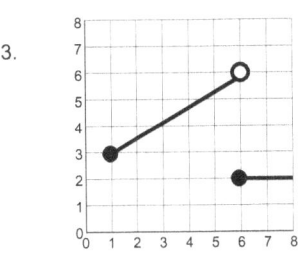 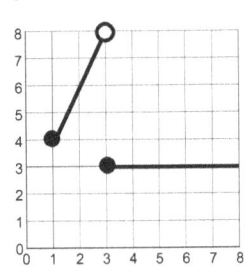

$f(x) = \begin{cases} \underline{\hspace{1cm}} \text{ if } \underline{\hspace{1cm}} \\ \underline{\hspace{1cm}} \text{ if } \underline{\hspace{1cm}} \end{cases}$ $f(x) = \begin{cases} \underline{\hspace{1cm}} \text{ if } \underline{\hspace{1cm}} \\ \underline{\hspace{1cm}} \text{ if } \underline{\hspace{1cm}} \end{cases}$

#3 Solve these story problems

1. You have a summer job that pays time and a half for overtime. It's 1.5 times your hourly rate of Rs 60/hr.

$$f(x) = \begin{cases} \underline{\hspace{2cm}} & \text{if } \underline{\hspace{2cm}} \\ \underline{\hspace{2cm}} & \text{if } \underline{\hspace{2cm}} \end{cases}$$

Calculator? yes   no

2. Electricity is Rs 30/kwh for the first 1,000 kwh and then drops to Rs 10/kwh for usage over 1,000 kwh.

$$f(x) = \begin{cases} \underline{\hspace{2cm}} & \text{if } \underline{\hspace{2cm}} \\ \underline{\hspace{2cm}} & \text{if } \underline{\hspace{2cm}} \end{cases}$$

3. The Rental Store charge for a drill is Rs 600 for 1 day and it's Rs 400 for each additional day after that.

$$f(x) = \begin{cases} \underline{\hspace{2cm}} & \text{if } \underline{\hspace{2cm}} \\ \underline{\hspace{2cm}} & \text{if } \underline{\hspace{2cm}} \end{cases}$$

4. A national fraternity allows 1 delegate for each 20 members or less and 2 delegates for larger than 20 members.

$$f(x) = \begin{cases} \underline{\hspace{2cm}} & \text{if } \underline{\hspace{2cm}} \\ \underline{\hspace{2cm}} & \text{if } \underline{\hspace{2cm}} \end{cases}$$

5. A salesperson receives a Rs 2000 bonus for each Rs 5,000 in sales and Rs 1000 if they have over Rs 10,000 in sales per week.

$$f(x) = \begin{cases} \underline{\hspace{2cm}} & \text{if } \underline{\hspace{2cm}} \\ \underline{\hspace{2cm}} & \text{if } \underline{\hspace{2cm}} \end{cases}$$

6. A taxi company charges Rs 150 per km for trips less than 10K and Rs 80 for each km over 10 km.

$$f(x) = \begin{cases} \underline{\hspace{2cm}} & \text{if } \underline{\hspace{2cm}} \\ \underline{\hspace{2cm}} & \text{if } \underline{\hspace{2cm}} \end{cases}$$

7. An economy car costs Rs 8100 for 1 week. They also charge Rs 1500 per day each day over the week.

$$f(x) = \begin{cases} \underline{\hspace{2cm}} & \text{if } \underline{\hspace{2cm}} \\ \underline{\hspace{2cm}} & \text{if } \underline{\hspace{2cm}} \end{cases}$$

8. In 2015, the gas company charged monthly customers Rs 70 for using less than 20 therms at Rs 20/thermand. If they use over 300 therms, they charge Rs 150.

$$f(x) = \begin{cases} \underline{\hspace{2cm}} & \text{if } \underline{\hspace{2cm}} \\ \underline{\hspace{2cm}} & \text{if } \underline{\hspace{2cm}} \end{cases}$$

9. A credit card company charges Rs 100 for bills of Rs 1000 or less and there is a Rs 30 charge for customers over Rs 1000.

$$f(x) = \begin{cases} \underline{\hspace{2cm}} & \text{if } \underline{\hspace{2cm}} \\ \underline{\hspace{2cm}} & \text{if } \underline{\hspace{2cm}} \end{cases}$$

Ch 12 Ls 1 Substitute simultaneous equations.   119

_____ #1 #2 ____/ 7   #3 ____/ 8   R ____/ 7   Total ____/ 22 _____
        Name   Checker

#1 1. What do you need to substitute equations? _____
     2. What happens next? _____
     3. What does the answer find? _____

         4. What substitutes into what? $y = 2x + 1$
                                                 $x + y = 6$

                                                                      Make the
                                         _____ subs in for _____ **equation.**

         Solve the equation.   _____

         How do you find Y?   _____

         Solve for Y.   _____

         The point is _____.   _____

#2 1. When is there a new first step? _____
     2. What happens to get a variable by itself? _____

         3. What substitutes into what? $y = x + 5$
                                                $2y = 4x - 2$

         Make the equation.   _____ subs in for _____

         Solve the equation.   _____

         How do you find Y?   _____

         Solve for Y.   _____

         The point is _____.   _____

#3 What can you substitute? Find the point.   Calculator?
                                                    yes   no

1. $3x + 2 = y$ $a + 2b = 11$
     $x + y = 8$ $4a + 1 = b$

Pick an
equation.

                  (,)   (,)

2. $3x + 1 = y$ $a + 2b = 16$
     $x + y = 9$ $3a + 1 = b$

Pick an
equation.

                  (,)   (,)

Review 1. What do you need to substitute equations? _____   Calculator?
        2. Solve for a variable. What happens next? _____   yes   no
        3. What does the answer find? _____
        4. When is there a new first step? _____
        5. What happens to get a variable by itself? _____

Substitute into   6. $2x + 1 = y$ $a + 3b = 8$
the other equation. $3x + y = 12$ $2a + 1 = b$

Ch 12 Ls 2 Eliminate simultaneous equations.   121

_____ #1 #2 ____/ 7   #3 ____/ 8 R ____/ 9 Total ____/ 24 _____
          Name   Checker

#1 1. What do you need to eliminate a variable? _____
    2. How do you eliminate a variable? _____
    3. What step subtracts an equation? _____
    4. How do you change an equation to eliminate a variable? _____
    5. You know X. How do you find Y? _____

    #2   1. What eliminates what? $3x + 2y = 4$
                                           $-3x + y = 11$

              Solve the equation.   _____

              How do you find Y?   _____

                                           _____
              Solve for Y.   _____

              The point is _____.   _____

         2. What changes so you $2x - 5y = 1$
            can eliminate it? $-x + 2y = 4$ _____

              Make the equation.   _____

              Solve the equation.   _____

              How do you find Y?   _____

              Solve for Y.   _____

              The point is _____.   _____

#3   What can you eliminate? Find the point.   Calculator? yes   no

1. $2x + y = 2$
    $x - y = 4$

    $a + 3b = 8$
    $-a + 2b = 7$

Pick an equation. _____

_____

_____

_____ (,)

_____

_____

_____

_____ (,)

2. $3x + 2y = 2$
    $x + 2y = 4$

    $4a + 5b = 9$
    $-4a + b = 3$

Pick an equation. _____

_____

_____

_____ (,)

_____

_____

_____

_____ (,)

Review
1. What do you need to eliminate a variable? _____
2. How do you eliminate a variable? _____
3. What step subtracts an equation? _____
4. How do you change an equation to eliminate a variable? _____
5. You know X. How do you find Y? _____

Calculator? yes   no

6. $2x + y = 1$
    $3x + 2y = 4$ _____

    $a + 3b = 5$
    $2a + 2b = 2$ _____

Change 1 equation so they combine.

7. $2x + y = 2$
    $x + 3y = 6$ _____

    $a + b = 5$
    $4a + 2b = 4$ _____

Ch 12 Ls 3 Simultaneous equation story problems.   123

_____ #1 #2 ____ / 5   #3 ____ / 4   R ____ / 5   T ____ / 14 _____
          Name   Checker

#1 1. What does it mean by Same Equation problems? _____
    2. Why would they be called Different Equations? _____
    3. What do you look at to decide which it is? _____

#2 1. TJ mixed Rs 160 raisins/kg mixes with nuts worth Rs 245/kg
    to make 17 kg worth Rs 200/ kg. How many kilograms of raisins   What are the
    and nuts should she use? Use 2 equation answers.   2 equations?

            First equation   _____
                                                        Combine equations.
            2nd equation   _____

            Solve the equation.   _____

            Find the other variable. _____

                                  _____

                                  _____   (___, ___)

    2. You sold soccer tickets for students for Rs 50 and adults   What are the
       are Rs 80. They sold Rs 53,600. How many of each are sold?   2 equations?

            First equation   _____
                                                        Combine equations.
            2nd equation   _____

            Solve the equation.   _____

            Find the other variable. _____

                                  _____

                                  _____   (___, ___)

#3 1. A total of Rs 13,000 is invested in 2 funds at 5% and 8% for Rs 890 simple interest. How much is invested in each account?

Calculator? yes no

_____

_____

1st _____

_____

2nd _____

_____

2. A total of R 15,000 is invested in 2 funds at 6% and 8% for R 1,100 simple interest. How much is invested in each account?

_____

1st _____

_____

2nd _____

_____

3. Mrs D spends some time training in an airplane and in a simulator. He's required to have 60 hrs. Time in a simulator is Rs 3000 per hour and in a plane is Rs 9000. She has Rs 200,000 for her pilot license.

_____

_____

1st _____

_____

2nd _____

_____

Review 1. What does it mean by Same Equation problems? _____

Calculator? yes no

2. Why would they be called Different Equations? _____

3. What do you look at to decide which it is? _____

Story Problems. 125

_____ #1 ____ / 4   #2 ____ / 4   Total ____ / 8
Name

#1 1. Pari mixes candy that sells for Rs 30/kg with   1st   _____   Calculator?
    candy that costs Rs 40/lb to make 40 kg of   yes   no
    candy selling for Rs 370/lb. How many kilo-   2nd   _____
    grams of each kind of candy did she use in   _____
    the mix?   _____

Find both equations,   _____
    then solve.   _____
                                                            _____
                            (____ , ____)   _____

    2. Tickets to a concert cost either Rs 80 or Rs   1st _____
    120. A total of 200 tickets are sold, and the   _____
    total receipts were Rs 2200. How many of   2nd   _____
    each kind of ticket were sold?   _____
                                                            _____
                                                            _____
                                                            _____
                            (____ , ____)   _____

    3. A coffee store sold 62 teas and coffees for   1st _____
    a total sales of Rs 2900. Tea is Rs 40 and   _____
    coffee is Rs 50 each. How many of each   2nd   _____
    were sold?   _____
                                                            _____
                                                            _____
                                                            _____
                            (____ , ____)   _____

    4. An investor buys a total of 400 shares of two   1st _____
    stocks. The price of one stock is Rs 200 per   _____
    share, while the price of the other stock is   2nd   _____
    Rs 500 per share. The investor spends a   _____
    total of Rs 80,000.   _____

    How many shares of each stock did the   _____
    investor buy?   _____
                            (____ , ____)   _____

#2

1. Ray mixes an alloy containing 20% silver with an alloy containing 40% silver to make 80 kg of an alloy with 25% silver. How many pounds of each kind of alloy did he use?

   1st _____
   2nd _____

   Calculator? yes   no

   (____, ____)

2. The sum of two numbers is 90. The larger number is 14 more than 3 times the smaller number. Find the numbers.

   1st _____
   2nd _____

   (____, ____)

3. Benassi mixes an alloy containing 15% silver with an alloy containing 25% silver to make 60 kgs of an alloy with 18% silver. How many kilograms of each kind of alloy did she use?

   1st _____
   2nd _____

   (____, ____)

4. How many liters of each of a 50% acid solution and an 70% acid solution must be mixed to produce 50 liters of a 64% acid solution?

   1st _____
   2nd _____

   (____, ____)

Review Problems. 127

_____ #1 #2 ____/ 10   #3 ____/ 12   Total ____/ 30
           Name

#1 1. Simultaneous Equations _____
    2. Substitute Equatiions _____
    3. Elimination Equations _____
    4. Equal Equations _____
    5. Same Equations _____
    6. Different Equations _____

#2 Substitute or eliminate? Find the point.   Calculator?
                                                      yes   no

1. $-4x + 1 = y$ $a + 3b = 9$
     $x + y = 2$ $3a + 3b = 2$
     _____   _____
     _____   _____
Pick an
equation. _____   _____
     _____   _____
     _____ (,)   _____ (,)

2. $5x - 3 = y$ $a + 2b = 14$
     $x + 2 = y$ $7a - 3 = b$
     _____   _____
     _____   _____
Pick an
equation. _____   _____
     _____   _____
     _____ (,)   _____ (,)

#3 Substitute or eliminate? Find the point.

Calculator? yes no

1.
$a + 3b = 9$
$5a - 2 = b$

Substitute ___
Eliminate ___

Pick an equation. ___

(,)

$4x - 1 = y$
$x - 6 = y$

Substitute ___
Eliminate ___

(,)

2.
$a + 4b = 10$
$2a + 4b = 3$

Substitute ___
Eliminate ___

Pick an equation. ___

(,)

$-5x + 2 = y$
$x + y = 4$

Substitute ___
Eliminate ___

(,)

3.
$-4x + 6 = y$
$x + y = 8$

Substitute ___
Eliminate ___

Pick an equation. ___

(,)

$a + 5b = 12$
$3a + 2b = 7$

Substitute ___
Eliminate ___

(,)

Quadratics

The 7th course in the Math Without Calculators Courses.

Ch 1 Ls 1 What begins quadratics. 129

_____ #1 #2 ____/ 18 #3 ____/ 20 R ____/ 12 T ____/ 50 _____
         Name   Checker

#1 1. What makes a quadratic variable? _____
2. If a quadratic solves a negative number, what sign does it make? _____
3. How does a quadratic change an answer? _____
4. What happens if a negative squared variable solves a positive? _____
5. Square 0.2. Find the answer. _____   Square 1 third. What's the answer? _____
6. Why does a fraction use parentheses to square it? _____
7. What shape multiplies same variables? _____
8. What shape multiplies a binomial? _____

#2 1 Solve these squared numbers. 3^2 $(-3)^2$

_____   _____

2. Solve with negatives. $-(3^2)$ $-(-3)^2$

_____   _____

3. How does a fraction get squared? $\left(\frac{1}{4}\right)^2$ $\left(-\frac{2}{3}\right)^2$

_____   _____

4. How does a decimal multiply? 0.2^2 1.4^2

_____   _____

5. How does a decimal multiply? 0.01^2 0.04^2

_____   _____

#3 Solve these fractions and decimals   Calculator?
                                                                      yes   no

1. $\left(\dfrac{1}{5}\right)^2$ $\left(-\dfrac{1}{4}\right)^2$ $-\left(\dfrac{1}{2}\right)^2$ $\left(\dfrac{3}{5}\right)^2$ $-\left(\dfrac{1}{7}\right)^2$

   _____   _____   _____   _____   _____

2. $-\left(\dfrac{1}{6}\right)^2$ $\left(\dfrac{2}{5}\right)^2$ $\left(\dfrac{1}{3}\right)^2$ $-\left(\dfrac{1}{8}\right)^2$ $\left(-\dfrac{3}{4}\right)^2$

   _____   _____   _____   _____   _____

3. 0.8^2 0.05^2 0.7^2 1.2^2 1.02^2

   _____   _____   _____   _____   _____

4. 0.1^2 0.3^2 0.04^2 0.6^2 0.12^2

   _____   _____   _____   _____   _____

Review 1. What makes a quadratic variable? _____   Calculator?
2. If a quadratic solves a negative number, what sign does it make? _____   yes   no
3. How does a quadratic change an answer? _____
4. What happens if a negative squared variable solves a positive? _____
5. Square 0.2. Find the answer. _____   Square 1 third. What's the answer? _____
6. Why does a fraction use parentheses to square it? _____
7. What shape multiplies same variables? _____
8. What shape multiplies a binomial? _____

9.
Find the   x [   ]   _____   x [   ]   _____
area   x + 3   x + 6
if x is 4.

10.
              x [   ]   _____   x [   ]   _____
                 x + 5   x + 4

                                                           Each lesson has a quiz.

Ch 1 Ls 2 Graph a quadratic with a y intercept.   131

_____ #1 #2 ____ / 11 #3 ____ / 12 R ___ / 8 T ____ / 31 _____
Name   Checker

#1 1. Name 2 things about a quadratic graph. _____

2. What turns a graph down? _____

3. Why do quadratic graphs make a cup? _____

4. What does the C term show? _____

5. How can you decide if an equation has X intercepts? _____

6. What is the center line of a parabola called? _____

7. What point does the quadratic change direction? _____

#2 1. Solve for X is 2 and -2. $y = -x^2 + 3$

Predict a graph. X intercepts?   When x is 2 it's _____   When x is -2 it's _____.

Opens Up   Down
Y intercept ____

X intercepts
yes   no

2. Solve for X is 2 and -2. $y = x^2 - 2$

Predict a graph. X intercepts?   When x is 2 it's _____   When x is -2 it's _____.

Opens Up   Down
Y intercept ____

X intercepts
yes   no

#3

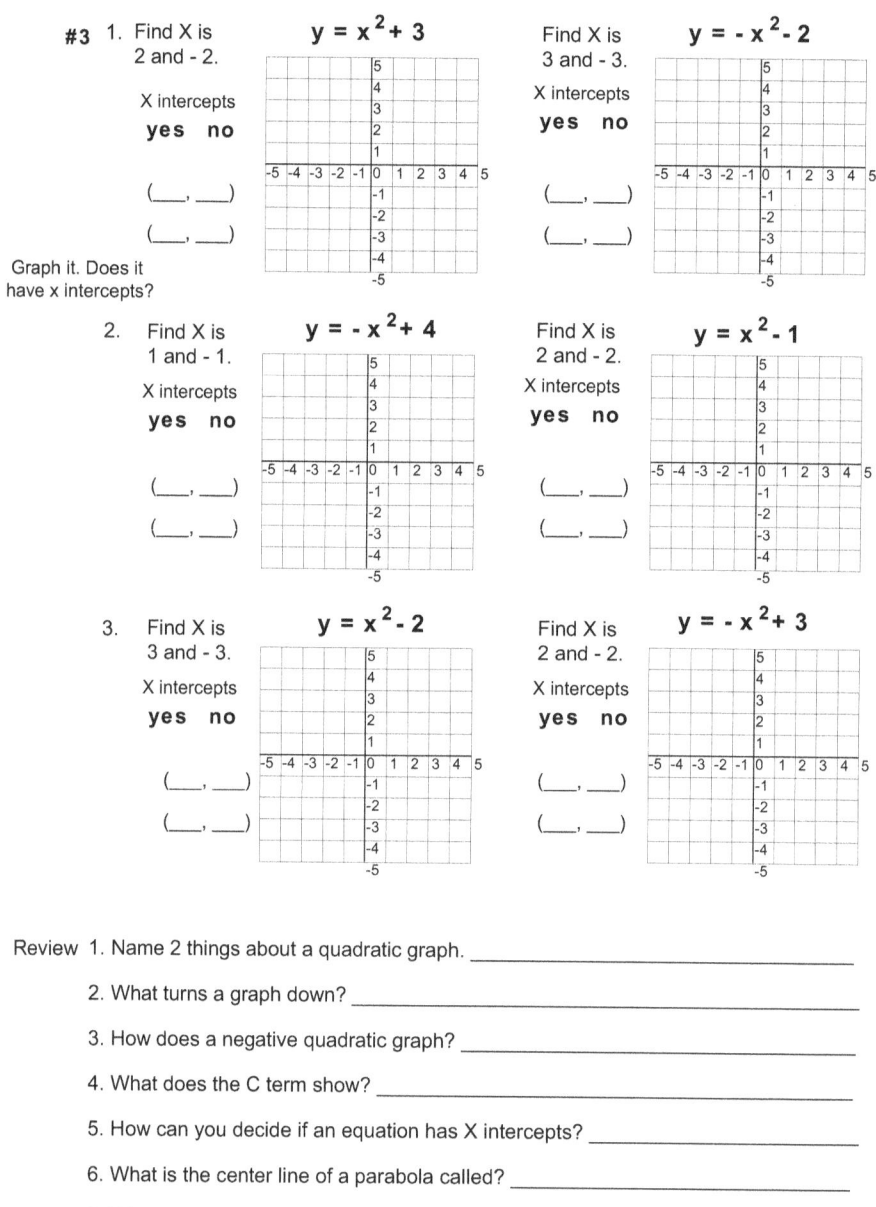

1. Find X is 2 and -2. $y = x^2 + 3$ Find X is 3 and -3. $y = -x^2 - 2$ Calculator? yes no

   X intercepts yes no

   X intercepts yes no

   (___, ___)
   (___, ___)

   (___, ___)
   (___, ___)

   Graph it. Does it have x intercepts?

2. Find X is 1 and -1. $y = -x^2 + 4$ Find X is 2 and -2. $y = x^2 - 1$

   X intercepts yes no

   X intercepts yes no

   (___, ___)
   (___, ___)

   (___, ___)
   (___, ___)

3. Find X is 3 and -3. $y = x^2 - 2$ Find X is 2 and -2. $y = -x^2 + 3$

   X intercepts yes no

   X intercepts yes no

   (___, ___)
   (___, ___)

   (___, ___)
   (___, ___)

Review 1. Name 2 things about a quadratic graph. _____

2. What turns a graph down? _____

3. How does a negative quadratic graph? _____

4. What does the C term show? _____

5. How can you decide if an equation has X intercepts? _____

6. What is the center line of a parabola called? _____

7. What point does the quadratic change direction? _____

Ch 1 Ls 3 How A term changes quadratics.   133

_____ #1 #2 ____ / 10   #3 ____ / 12   R ___/ 4   T ____/ 26 _____
Name   Checker

#1 1. Name 2 steps to multiply $2x^2$ for x is - 2. _____

2. How does A term change a standard graph? _____

3. Name 3 steps to predict a parabola. _____

_____

4. What does SOY X show you? _____

#2   1. Solve for X is 2 and - 2. $y = -2x^2 + 3$

Predict a graph. X intercepts?   When x is 2 it's _____   When x is - 2 it's _____.

Opens Up Down
Y intercept ___
X intercepts
yes   no

2. Solve for X is 2 and - 2. $y = 0.5x^2 - 2$

Predict a graph. X intercepts?   When x is 2 it's _____   When x is - 2 it's _____.

Opens Up Down
Y intercept ___
X intercepts
yes   no

134

#3

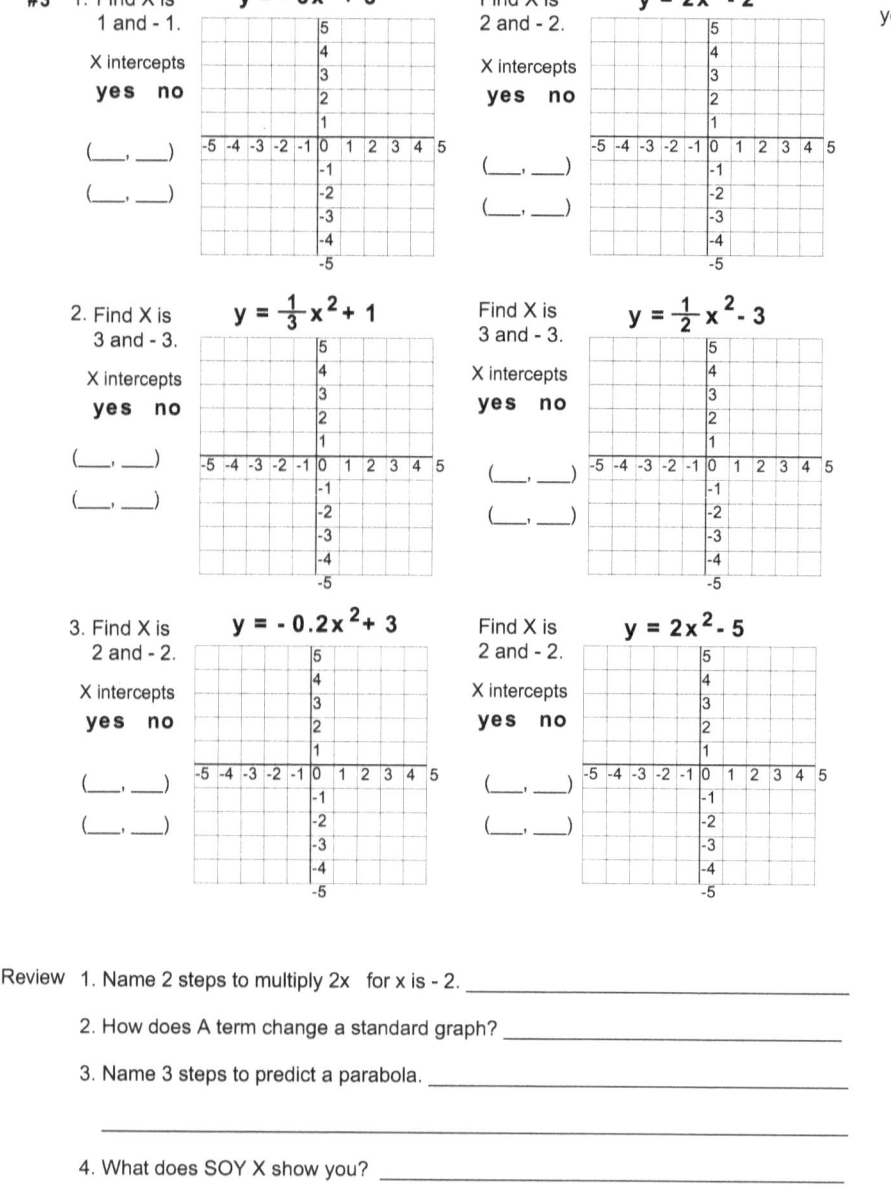

1. Find X is 1 and -1. $y = -3x^2 + 3$ Find X is 2 and -2. $y = 2x^2 - 2$ Calculator? yes no

   X intercepts yes no

   (__, __)
   (__, __)

   X intercepts yes no

   (__, __)
   (__, __)

2. Find X is 3 and -3. $y = \frac{1}{3}x^2 + 1$ Find X is 3 and -3. $y = \frac{1}{2}x^2 - 3$

   X intercepts yes no

   (__, __)
   (__, __)

   X intercepts yes no

   (__, __)
   (__, __)

3. Find X is 2 and -2. $y = -0.2x^2 + 3$ Find X is 2 and -2. $y = 2x^2 - 5$

   X intercepts yes no

   (__, __)
   (__, __)

   X intercepts yes no

   (__, __)
   (__, __)

Review 1. Name 2 steps to multiply 2x for x is - 2. _____

2. How does A term change a standard graph? _____

3. Name 3 steps to predict a parabola. _____
_____

4. What does SOY X show you? _____

Ch 1 Ls 4 Solve trinomials for points.   135

_____ #1 #2 ____ / 8 #3 ____ / 14 R ____ / 3 T ____ / 25 _____
           Name   Checker

#1 1. Name the 1st step to solve $2x^2 + 3x + 1$ for x is 2? _____
    2. What are the next 2 steps? _____
    3. After you solve the 2 steps, what does it find? _____

#2 1. Find X is 3. What 3 numbers add? $y = x^2 + 2x + 1$

              When x is 3 it adds ____ + ____ + ____   What's the point?

                        Point is 3, ____

    2. Solve for X is 2. Find 3 numbers. $y = x^2 + 3x + 5$

              When x is 2 it adds ____ + ____ + ____   What's the point?

                        Point is 2, ____

    3. Solve for X is 5. Find 3 numbers. $y = 2x^2 + 4x + 3$

              When x is 3 it adds ____ + ____ + ____   What's the point?

                        Point is 5, ____

    4. Solve for X is 4. Find 3 numbers. $y = 2x^2 + 4x + 1$

              When x is 4 it adds ____ + ____ + ____   What's the point?

                        Point is 4, ____

    5. Solve for X is 10. Find 3 numbers. $y = \frac{1}{2}x^2 + 4x + 2$

              When x is 10 it adds ____ + ____ + ____   What's the point?

                        Point is 10, ____

#3 What's the answer? Calculator? yes no

1. $y = x^2 + 5x + 1$ $y = x^2 + 3x + 7$
   Point is 2, ___   Point is 4, ___

2. $y = \frac{1}{2}x^2 + 4x + 2$ $y = 2x^2 + 2x + 1$
   Point is 2, ___   Point is 4, ___

3. $y = 2x^2 + 5x + 4$ $y = 4x^2 + 6x + 1$
   Point is 3, ___   Point is 5, ___

4. $y = 2x^2 + 2x + 1$ $y = \frac{1}{2}x^2 + 4x + 2$
   Point is 4, ___   Point is 2, ___

5. $y = \frac{1}{2}x^2 + 3x + 3$ $y = \frac{1}{2}x^2 + 3x + 6$
   Point is 4, ___   Point is 3, ___

6. $y = 4x^2 + 6x + 4$ $y = 2x^2 + 5x + 8$
   Point is 3, ___   Point is 2, ___

7. $y = \frac{1}{3}x^2 + 4x + 1$ $y = \frac{1}{4}x^2 + 3x + 2$
   Point is 3, ___   Point is 2, ___

Review 1. Name the 1st step to solve $2x^2$ for x is -2? ___
2. What are the next 2 steps? ___
3. After you solve the 1st step, what does it find? ___

Review Problems 137

_____ #1 #2 ____ / 11 #3 #4 ____ / 2 R ___ / 9 T ____ / 22 _____
Name   Checker

#1 1. Quadratics _____

     2. C Term _____

     3. A term _____

#2 Solve these fractions and decimals   Calculator?
                                                                 yes   no

1. $-\left(\dfrac{1}{8}\right)^2$ $\left(\dfrac{3}{7}\right)^2$ $\left(\dfrac{5}{6}\right)^2$ $-\left(\dfrac{2}{3}\right)^2$ $\left(\dfrac{-8}{9}\right)^2$

    _____   _____   _____   _____   _____

2. 0.4^2 0.03^2 0.6^2 2.5^2 0.05^2

    _____   _____   _____   _____   _____

#3 1. Find X is $y = -x^2 + 2$ Find X is $y = x^2 - 3$ Calculator?
       2 and -2.   3 and -3.   yes   no
       X intercepts   X intercepts
       yes   no   yes   no

       (__, __)   (__, __)
       (__, __)   (__, __)

    2. Find X is $y = -x^2 + 1$ Find X is $y = x^2 - 4$
       2 and -2.   1 and -1.
       X intercepts   X intercepts
       yes   no   yes   no

       (__, __)   (__, __)
       (__, __)   (__, __)

#4 How does slope change a parabola?   Calculator? yes no

1. Find X is 2 and -2. $y = -2x^2 + 1$ Find X is 1 and -1. $y = 3x^2 - 3$

   X intercepts yes no

   (___, ___)
   (___, ___)

   X intercepts yes no

   (___, ___)
   (___, ___)

2. Find X is 2 and -2. $y = \frac{1}{4}x^2 + 1$ Find X is 2 and -2. $y = \frac{1}{2}x^2 - 2$

   X intercepts yes no

   (___, ___)
   (___, ___)

   X intercepts yes no

   (___, ___)
   (___, ___)

#5 Solve these quadratics for the point.   Calculator? yes no

1. $y = x^2 + 3x + 2$ $y = x^2 + 2x + 9$

   Point is 2, _____   Point is 4, _____

2. $y = \frac{1}{3}x^2 + 5x + 4$ $y = 2x^2 + 3x + 5$

   Point is 2, _____   Point is 4, _____

3. $y = 3x^2 + 4x + 1$ $y = \frac{1}{5}x^2 + 2x + 3$

   Point is 3, _____   Point is 5, _____

4. $y = \frac{1}{2}x^2 + 5x + 2$ $y = 4x^2 + 6x + 1$

   Point is 2, _____   Point is 3, _____

Ch 2 Ls 1 How squares change signs. 139

_____ #1 #2 ____/ 9 #3 ____/ 9 R ____/ 11 T ____/ 29 _____
           Name   Checker

#1 1. How does x^2 solve for - 2? _____   How does 3x solve for - 2? _____

2. Name 2 things that decide terms for A and B. _____

3. Name 3 steps to solve a trinomial with an A term. _____

_____

4. Solve for X is 2. Which term is bigger? How much? $2x^2 + 3x$

_____ is bigger by _____

5. Solve for X is 2. Which is bigger? How much? $\frac{1}{2}x^2 - 3x$

_____ is bigger by _____

6. Solve for X is 2. Which is bigger? How much? $0.7x^2 \quad 2x$

_____ is bigger by _____

#2 1. Find X is 3. What 3 numbers add? $y = x^2 - 2x - 1$

When x is 3 it adds ____ - ____ - ____   What's the point?

Point is 3, ____

2. Solve for X is - 2. Find 3 numbers. $y = 2x^2 + 4x + 3$

When x is - 2 it adds ____ - ____ + ____   What's the point?

Point is - 2, ____

3. Solve for X is 4. Find 3 numbers. $y = \frac{1}{2}x^2 + 4x + 2$

When x is 4 it adds ____ + ____ + ____   What's the point?

Point is 4, ____

#3 Find which number is greater? Solve for x is 2.   Calculator? yes   no

1. $-2x^2 - 2x$ $-3x^2 - 4x$ $-x^2 - 3x$

Circle the larger answer.   ___ ___   ___ ___   ___ ___

2. $\frac{1}{2}x^2 - x$ $-\frac{1}{2}x^2 + 2x$ $\frac{1}{2}x^2 \quad 4x$

___ ___   ___ ___   ___ ___

3. $3x^2 + 6x$ $4x^2 + 2x$ $6x^2 + 3x$

___ ___   ___ ___   ___ ___

Review 1. How does x^2 solve for -2? _____ How does $3x$ solve for -2? _____   Calculator? yes   no

2. Name 2 things that decide terms for A and B. _____

3. Name 3 steps to solve a trinomial with an A term. _____

_____

Solve these quadratics.

4. $y = 2x^2 - 2x + 1$ $y = \frac{1}{2}x^2 - 4x - 2$

When x is 2 (___, ___)   When x is -2 (___, ___)

5. $y = x^2 - 4x + 2$ $y = 3x^2 - x - 2$

When x is 5 (___, ___)   When x is -2 (___, ___)

6. $y = x^2 - 2x + 6$ $y = \frac{1}{2}x^2 - x - 2$

When x is 3 (___, ___)   When x is -1 (___, ___)

7. $y = 3x^2 + 2x - 3$ $y = \frac{1}{3}x^2 + x + 3$

When x is -3 (___, ___)   When x is 3 (___, ___)

Ch 2 Ls 2 Find axis of symmetry and vertex. 141

_____ #1 #2 ____/ 9 #3 ____/ 12 R ___/ 5 T ____/ 22 _____
           Name   Checker

#1 1. How does B term show vertex side? _____
     2. With Axis of Symmetry, where do you start to find the axis? _____
     3. With Axis of symmetry, what does - b divide by? _____
     4. What formula finds the Axis of Symmetry? _____
     5. After you find the axis (X), what finds the vertex? _____

#2 1. With Axis of Symmetry, what is the 1st step to find the axis? $y = x^2 + 2x$

    What is the 2nd step to find the axis? Negative B is _____

    Solve it. Where's the axis of symmetry? Divide by _____

                                                Equals _____

  2. What is the 1st step to find the axis? $y = 2x^2 - 4x$

    What is the 2nd step to find the axis? Negative B is _____

    Solve it. Where's the axis of symmetry? Divide by _____

                                               Equals _____

  3.   In 1 step, what's the axis? $y = x^2 - 8x$

    Solve it. Where's the vertex? _____

                                               Equals _____

  4.   In 1 step, what's the axis? $y = 2x^2 - 8x$

    Solve it. Where's the axis of symmetry? _____

                                               Equals _____

#3 1. $y = x^2 + 4x + 1$ $y = 2x^2 - 2x - 1$ Calculator?
                                                                                                      yes   no

(2,   )   (2,   )
(- 2,)   (- 2,)

Opens ____   Opens ____
Y int ____   Y int ____
Vertex   Vertex
(____, ____)   (____, ____)

2. $y = x^2 - 6x + 3$ $y = x^2 + 2x - 2$

(2,   )   (2,   )
(- 2,)   (- 2,)

Opens ____   Opens ____
Y int ____   Y int ____
Vertex   Vertex
(____, ____)   (____, ____)

3. $y = 2x^2 + 8x - 3$ $y = 2x^2 - 4x - 1$

(2,   )   (2,   )
(- 2,)   (- 2,)

Opens ____   Opens ____
Y int ____   Y int ____
Vertex   Vertex
(____, ____)   (____, ____)

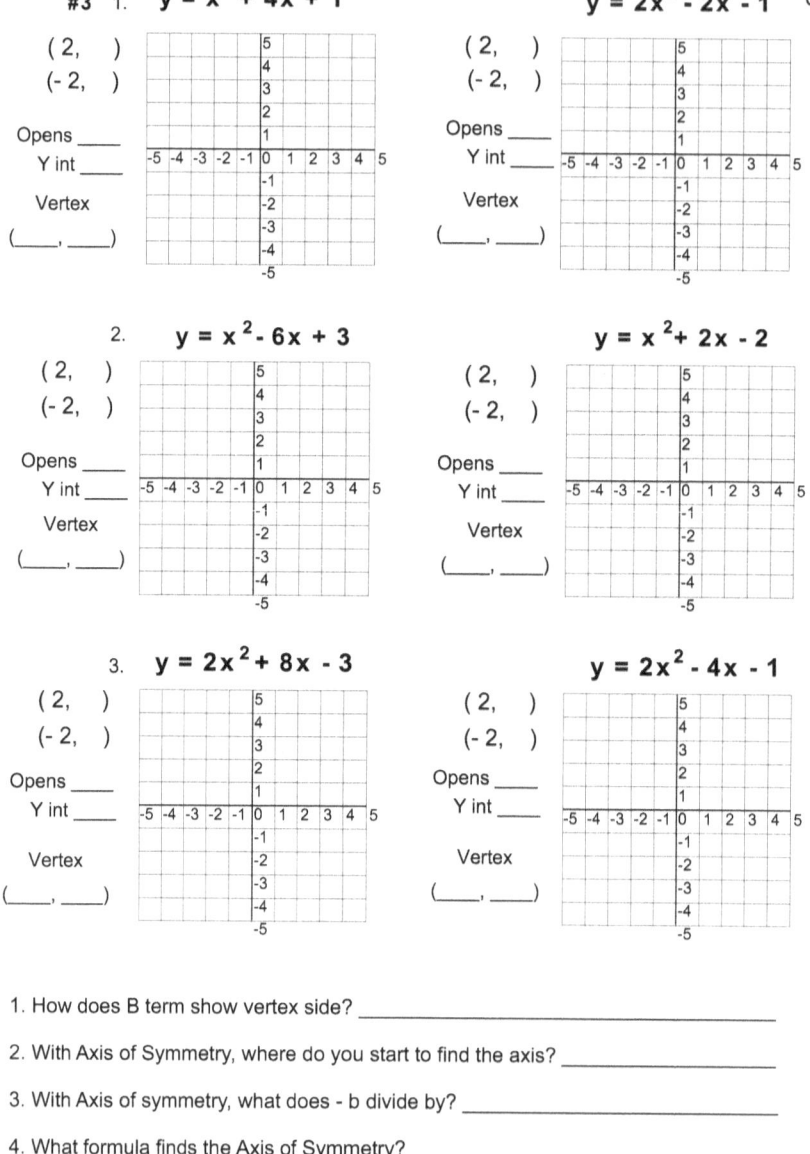

Review 1. How does B term show vertex side? _____

2. With Axis of Symmetry, where do you start to find the axis? _____

3. With Axis of symmetry, what does - b divide by? _____

4. What formula finds the Axis of Symmetry? _____

5. After you find the axis (Y), what finds the vertex? _____

Ch 2 Ls 3 How Axis of Symmetry makes points. 143

_____ #1 #2 ____/ 8 #3 ____/ 10 R ____/ 3 T ____/ 21 _____
          Name   Checker

#1 1. If A is 1, what does the axis formula divide by? _____

2. If A is 1, which B numbers make whole axis numbers? _____

3. If you want a positive axis, what sign will B be? _____

#2 1. What sign makes a negative answer? $y = x^2 \ ?x$

What A term makes 1 as an answer? $y = x^2 \ 2x$

Put it together. Where is the axis? $y = \_\_ x^2 \ x$

- ___ over ___ is ___ or ☐

2. What sign makes a positive answer? $y = x^2 \ ?x$

What A term makes 2 as an answer? $y = x^2 - 2x$

Put it together. Where is the axis? $y = \_\_ x^2 \ x$

- ___ over ___ is ___ or ☐

3. What if A and B term are the same? $y = 2x^2 + 2x$

- ___ over ___ is ___ or ☐

4. What if A term is bigger than B? $y = 4x^2 + x$

- ___ over ___ is ___ or ☐

5. What if A and B term are both negative? $y = -2x^2 - 4x$

- ___ over ___ is ___ or ☐

#3 Find 2 different answers to get the axis at - 3.   Calculator? yes no

1. $y = \_\_x^2 \_\_x$ $y = \_\_x^2 \_\_x$

__ over __ is __   __ over __ is __

Find 2 different answers to get the axis at + 1.

2. $y = \_\_x^2 \_\_x$ $y = \_\_x^2 \_\_x$

__ over __ is __   __ over __ is __

Find 2 different answers to get the axis at + 0.5.

3. $y = \_\_x^2 \_\_x$ $y = \_\_x^2 \_\_x$

__ over __ is __   __ over __ is __

Find 2 different answers to get the axis at + 2.

4. $y = \_\_x^2 \_\_x$ $y = \_\_x^2 \_\_x$

__ over __ is __   __ over __ is __

Find 2 different answers to get the axis at + 4.

5. $y = \_\_x^2 \_\_x$ $y = \_\_x^2 \_\_x$

__ over __ is __   __ over __ is __

Review 1. If A is 1, what does the axis formula divide by? _____

2. If A is 1, which B numbers make whole axis numbers? _____

3. If you want a positive axis, what sign will B be? _____

Review Problems 145

_____ #1 to #3 ____/ 21 #3 #4 ____/ 16 T ____/ 37
            Name

#1 1. B Term _____
    2. Axis of Symmetry _____
    3. Vertex _____
    4. Whole Point _____

#2 Solve these quadratics.   Calculator?
                                                               yes   no

1. $y = 4x^2 - 3x + 5$ $y = \frac{1}{3}x^2 - 5x - 3$

    When x is 3 (___, ___)   When x is -2 (___, ___)

2. $y = x^2 - 6x + 2$ $y = 4x^2 - 3x - 2$

    When x is 4 (___, ___)   When x is -3 (___, ___)

3. $y = x^2 - 3x + 7$ $y = \frac{4}{5}x^2 - x - 3$

    When x is 5 (___, ___)   When x is -2 (___, ___)

4. $y = 3x^2 + 4x - 4$ $y = \frac{1}{2}x^2 + 2x + 6$

    When x is -3 (___, ___)   When x is 5 (___, ___)

#3 Use the Axis of Symmetry to find the vertex. 4 points each   Calculator?
                                                                  yes   no

1. $y = x^2 + 5x - 2$ $y = 2x^2 - 3x - 1$

(2,)   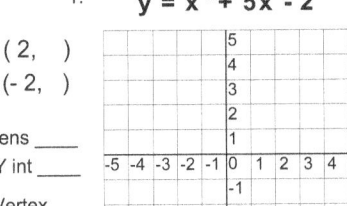   (2,)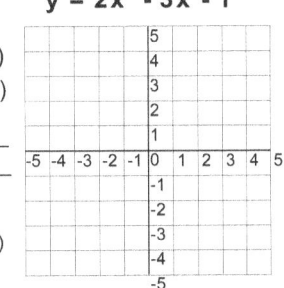
(-2,)   (-2,)

Opens ____   Opens ____
Y int ____   Y int ____
Vertex   Vertex
(____, ____)   (____, ____)

Part #3 continued.

2. $y = x^2 - 5x - 1$

(2, ___)
(-2, ___)
Opens ___
Y int ___
Vertex
(___ , ___)

$y = 2x^2 + 3x - 4$

(2, ___)
(-2, ___)
Opens ___
Y int ___
Vertex
(___ , ___)

#4 Find 2 different answers to get the axis at -2.   Calculator? yes   no

1. $y = \_\_\_x^2 \_\_\_x$ $y = \_\_\_x^2 \_\_\_x$

___ over ___ is ___   ___ over ___ is ___

Find 2 different answers to get the axis at + 2.

2. $y = \_\_\_x^2 \_\_\_x$ $y = \_\_\_x^2 \_\_\_x$

___ over ___ is ___   ___ over ___ is ___

Find 2 different answers to get the axis at - 0.5.

3. $y = \_\_\_x^2 \_\_\_x$ $y = \_\_\_x^2 \_\_\_x$

___ over ___ is ___   ___ over ___ is ___

Find 2 different answers to get the axis at + 5.

4. $y = \_\_\_x^2 \_\_\_x$ $y = \_\_\_x^2 \_\_\_x$

___ over ___ is ___   ___ over ___ is ___

Ch 3 Ls 1 How to multiply binomials. 147

_____ #1 #2 ____/ 6 #3 ____/ 9 R ____/ 9 T ____/ 15 _____
           Name   Checker

#1 1. Name the 1st 2 steps to multiply binomials? _____

      2. Name the last steps to multiply them? _____

      3. What is the last, last step? _____

#2 1. Multiply the 1st step for these binomials. $(a + 2)(a + 7)$

              Multiply the next step.   Firsts _____

              Multiply the 3rd step.   Outsides _____

              Multiply the 4th step.   Insides _____

              Put it together. What is the last step?   Lasts _____

                                                                     _____ _____ _____

      2. Multiply the 1st step. $(b + 1)(b + 6)$

              Multiply the next step.   Firsts _____

              Multiply the 3rd step.   Outsides _____

              Multiply the 4th step.   Insides _____

              Put it together. What is the last step?   Lasts _____

                                                                     _____ _____ _____

      3. Multiply the 1st 2 steps. $(x + 2)(x + 8)$

              Multiply the next 2 steps.   _____ _____

              What's the last, last step?   _____ _____

                                                                _____ _____ _____

#3 Multiply the 1st 2 steps, next 2 steps, then finish it.   Calculator?
                                                              yes   no

1. (x + 1)(x + 5)   (a + 3)(a + 1)   (b + 7)(b + 2)

    ___ ___   ___ ___   ___ ___

    ___ ___   ___ ___   ___ ___

    _____   _____   _____

2. (c + 6)(c + 3)   (d + 2)(d + 7)   (e + 1)(e + 5)

    ___ ___   ___ ___   ___ ___

    ___ ___   ___ ___   ___ ___

    _____   _____   _____

3. (f + 3)(f + 1)   (g + 6)(g + 5)   (h + 2)(h + 2)

    ___ ___   ___ ___   ___ ___

    ___ ___   ___ ___   ___ ___

    _____   _____   _____

Review 1. Name the 1st 2 steps to multiply binomials? _____   Calculator?
       2. Name the last steps to multiply them? _____   yes   no
       3. What is the last, last step? _____

4. (x + 2)(x + 8)   (a + 2)(a + 5)   (b + 7)(b + 3)

    _____   _____   _____

5. (c + 3)(c + 4)   (d + 5)(d + 6)   (e + 3)(e + 2)

    _____   _____   _____

Ch 3 Ls 2 Multiply binomials with negatives. 149

_____ #1 #2 ____/9 #3 ____/9 R ___/14 T ____/32 _____
Name   Checker

#1 1. How do all negatives change the answer? _____
    2. (e + 2)(e - 1) Will the C term be positive or negative? _____
    3. (e - 2)(e - 1) Will the C term be positive or negative? _____

         4. Multiply the 1st 2 steps.   (x - 2)(x - 4)

              Multiply the next 2 steps.   ____ ____

              What's the last, last step?   ____ ____

                                            ____ ____ ____

         5. Multiply the 1st 2 steps.   (d + 2)(- d - 4)

              Multiply the next 2 steps.   ____ ____

              What's the last, last step?   ____ ____

                                            ____ ____ ____

#2 1. (d + 3)(d - 3) Will the B term be negative? _____
    2. What happens to the trinomial? _____

         3. Multiply the 4 steps.   (a + 5)(a - 5)

              What's the last, last step?   ____ ____ ____ ____

                                            ____ ____ ____

         4. Multiply the 4 steps.   (b - 4)(b + 4)

              What's the last, last step?   ____ ____ ____ ____

                                            ____ ____ ____

#3 Multiply the 4 steps, then finish it.   Calculator? yes no

1. (x + 1)(x - 5)   (a + 3)(a - 1)   (b + 7)(b - 2)

    ___ ___ ___ ___   ___ ___ ___ ___   ___ ___ ___ ___
    _____   _____   _____

2. (c - 6)(c + 3)   (d - 2)(d + 7)   (e - 1)(e + 5)

    ___ ___ ___ ___   ___ ___ ___ ___   ___ ___ ___ ___
    _____   _____   _____

3. (f + 3)(f - 3)   (g + 6)(g - 6)   (h - 2)(h - 2)

    ___ ___ ___ ___   ___ ___ ___ ___   ___ ___ ___ ___
    _____   _____   _____

Review 1. How do all negatives change the answer? _____   Calculator? yes no
2. (e + 2)(e - 1) Will the C term be positive or negative? _____
3. (e - 2)(e - 1) Will the C term be positive or negative? _____
4. (d + 3)(d - 3) Will the B term be negative? _____
5. What happens to the trinomial? _____

6. (x - 2)(x - 4)   (a - 3)(a - 4)   (b - 4)(b - 3)
    _____   _____   _____

7. (a + 3)(a - 5)   (b - 6)(b - 4)   (x - 3)(x + 7)
    _____   _____   _____

8. (x + 5)(x - 5)   (a - 7)(a - 7)   (b + 4)(b - 4)
    _____   _____   _____

Ch 3 Ls 3 Multiply binomials with A terms.   151

_____ #1 #2 ____/ 7   #3 ____/ 9   R ____/ 12   T ____/ 28 _____
          Name   Checker

#1 1. (2e + 2)(e - 1) What will the A term be? _____
    2. How does it change the B term? _____
    3. Which term does it not change? _____

    #2   1. Multiply the 1st 2 steps.   (2x - 2)(3x - 4)

          Multiply the next 2 steps.   _____ _____

          What's the last, last step?   _____ _____

                                             _____ _____ _____

          2. Multiply the 1st 2 steps.   (3x + 1)(3x - 2)

          Multiply the next 2 steps.   _____ _____

          What's the last, last step?   _____ _____

                                             _____ _____ _____

          3. Multiply the 1st 2 steps.   (- 2x - 2)(5x - 4)

          Multiply the next 2 steps.   _____ _____

          What's the last, last step?   _____ _____

                                             _____ _____ _____

          4. Multiply the 1st 2 steps.   (- 2x - 2)(- 3x - 4)

          Multiply the next 2 steps.   _____ _____

          What's the last, last step?   _____ _____

                                             _____ _____ _____

#3. Use the A terms to solve it.   Calculator? yes   no

1. $(2x - 1)(x - 6)$ $(-4a + 4)(a - 1)$ $(3b + 5)(-b - 2)$

2. $(-2c - 5)(-c - 3)$ $(6d - 2)(2d - 7)$ $(-4e - 1)(3e + 5)$

3. $(4f - 3)(2f - 3)$ $(2g + 6)(g - 7)$ $(3h + 2)(h - 4)$

Review

1. $(2e + 2)(e - 1)$ What will the A term be? _____   Calculator? yes   no
2. How does it change the B term? _____
3. Which term does it not change? _____

4. $(3x - 1)(-2x + 5)$ $(2a - 3)(3a - 1)$ $(-6b - 7)(-2b - 2)$

5. $(2a + 3)(2a - 4)$ $(b + 2)(3b - 3)$ $(4x + 5)(x - 6)$

6. $(-4x + 1)(x - 5)$ $(-5a - 3)(a - 1)$ $(7b - 7)(b - 2)$

Ch 3 Ls 4 Multiply binomials the other way.   153

_____ #1 #2 ____ / 10 #3 ____ / 18 R ____ / 8 T ____ / 36 _____
        Name   Checker

#1 1. **(a + 2)(a + 4)** What's the 2nd way to multiply? _____

   2. What's next after multiplying? _____

#2 1. Multiply the 1st step for these binomials.   **(7a + 3)(a - 5)**

              Multiply the next step.   _____

              What's the answer?   _____

                                                _____

       2. Multiply the 1st step.   **(2a + 6)(a + 4)**

              Multiply the next step.   _____

              What's the answer?   _____

                                                _____

       3. Multiply the 1st step.   **(2a + 6)(0.5a + 4)**

              Multiply the next step.   _____

              What's the answer?   _____

                                                _____

       4. Multiply the 1st step.   **(3a + 3)(0.5a + 6)**

              Multiply the next step.   _____

              What's the answer?   _____

                                                _____

#3 Multiply the 1st 2 steps, next 2 steps, then finish it.   Calculator? yes   no

1. $(x + 1)(2x - 5)$ $(3a - 3)(a + 1)$ $(b - 7)(4b + 2)$

2. $(5c + 6)(2c + 3)$ $(d - 2)(3d - 7)$ $(-7e + 1)(e + 5)$

3. $(2f + 3)(f - 4)$ $(5g - 6)(g - 5)$ $(h - 2)(3h + 4)$

Review 1. $(a + 2)(a + 4)$ What's the 2nd way to multiply? _____   Calculator? yes   no

2. What's next after multiplying? _____

Choose 1st or 2nd way to multiply.

3. $(2x + 1)(5x + 5)$ $(6a - 3)(2a - 1)$ $(5b - 7)(-b - 2)$

4. $(x - 5)(x + 5)$ $(4a - 3)(3a - 1)$ $(-2b - 7)(b + 2)$

5. $(z - 2)(z + 4)$ $(-y - 1)(5y - 3)$ $(-4c + 1)(c - 3)$

Review Problems

_____ #1 to #3 ___ / 27 #4 #5 ___/18 T ___/ 45
Name

#1 1. 4 Names _____
    2. Multiply Binomials _____
    3. The Other Way _____

#2 1. $(x + 6)(x + 8)$ $(a + 7)(a + 5)$ $(b + 1)(b + 3)$ Calculator?
                                                                        yes   no

Multply these
quadratics.

    2. $(x + 3)(x + 7)$ $(a + 2)(a + 8)$ $(d + 9)(d + 2)$

    3. $(c + 5)(c + 5)$ $(d + 3)(d + 7)$ $(e + 4)(e + 2)$

    4. $(x + 4)(x + 8)$ $(a + 3)(a + 9)$ $(b + 3)(b + 6)$

#3 How do negatives change a quadratic?   Calculator?
                                                                        yes   no

    1. $(a + 2)(a - 6)$ $(b - 8)(b - 3)$ $(x - 6)(x + 8)$

    2. $(x - 3)(x - 4)$ $(a + 4)(a - 8)$ $(b - 2)(b - 5)$

    3. $(a + 6)(a - 7)$ $(b - 6)(b - 8)$ $(x - 4)(x + 9)$

    4. $(x + 6)(x - 7)$ $(a - 9)(a + 8)$ $(b + 8)(b - 9)$

#4. Use the A terms to solve it.   Calculator? yes no

1. $(4a - 3)(3a - 5)$ $(b + 3)(5b - 6)$ $(4x + 2)(2x - 4)$

   _____   _____   _____

2. $(5x - 1)(-3x + 7)$ $(2a - 3)(4a - 6)$ $(-4b - 7)(-3b - 5)$

   _____   _____   _____

3. $(-3x + 2)(x - 5)$ $(-5a - 2)(4a - 3)$ $(5c - 6)(2c - 2)$

   _____   _____   _____

#5 Multiply the other way.   Calculator? yes no

1. $(x + 2)(3x - 4)$ $(3a - 3)(5a + 1)$ $(b - 8)(3b + 3)$

2. $(5c + 6)(3c - 5)$ $(2d - 2)(4d - 9)$ $(-6e + 2)(e + 8)$

3. $(2f + 6)(5f - 7)$ $(4g - 3)(2g - 5)$ $(5h - 5)(2h + 6)$

Ch 4 Ls 1 Turn factors into binomials.   157

_____ #1 #2 ____/ 10   #3 ____/ 15   R ____/ 11   T ____/ 35 _____
             Name   Checker

#1 1. What does it mean to factor something? _____

2. What are the factors for these terms? $-8x^2 + 16x$ _____

3. What's the common factor? $2x^2 + 4x$

What is left over?   ____ (?   ?)

                              ___ (___ ___)

4. What's the common factor? $-b^2 - b$

What is left over?   ____ (?   ?)

                              ___ (___ ___)

5. All 1 step, how does it factor? $2a^2 + 4$

                              ___ (___ ___)

#2 1. What question do you ask if a fraction has a binomial? _____

2. How do you factor a binomial? _____

3. If a variable is in the denominator, what do you solve for? _____

4. Factor the numerator and finish it. $\dfrac{2a + 4}{a + 2}$

Cross out and simplify. $\dfrac{}{a + 2} = \dfrac{}{}$

5. How do you factor all the terms? $\dfrac{4x^2 - 2x}{2}$

                              _____

#3 Factor the numerator, then simplify the expression.   Calculator?
                                                          yes   no

1. $3a + 6 =$ _____ $3b - 3 =$ _____ $6c^2 + 2c =$ _____

2. $5d - 10 =$ _____ $2b^2 - b =$ _____ $4c^2 + 6c =$ _____

3. $6g - 1.5 =$ _____ $4a^2 - 2a =$ _____ $7b^2 - b =$ _____

4. $\dfrac{6x^2 + 2}{3x^2 + 1}$ $\dfrac{9x^2 - 3x}{6}$ $\dfrac{5a - 10}{a - 2}$

   _____ = ___   _____ = ___   _____ = ___

5. $\dfrac{3x^2 + x}{3x}$ $\dfrac{2x^2 - 5x}{x}$ $\dfrac{2a}{6a - 2}$

   _____ = ___   _____ = ___   _____ = ___

Review 1. What does it mean to factor something? _____   Calculator?
2. What are the factors for these terms? $-8x^2 + 16x$ _____   yes   no
3. What question do you ask if a fraction has a binomial? _____
4. How do you factor a binomial? _____
5. If a variable is in the denominator, what do you solve for? _____

6. $\dfrac{8}{4a + 2}$ $\dfrac{-b - 2}{b + 2}$ $\dfrac{a}{a^2 - 2a}$

   _____ = ___   _____ = ___   _____ = ___

7. $\dfrac{2c^2 + c}{c + 1}$ $\dfrac{2a^2 - 4a}{a - 2}$ $\dfrac{3a - 6}{6}$

   _____ = ___   _____ = ___   _____ = ___

Ch 4 Ls 2 Factor trinomial equations. 159

_____ #1 #2 ____/ 9 #3 ____/ 12 R ___/ 10 T ____/ 31 _____
            Name   Checker

#1 1. $x^2 + 6x + 5$ What's the 1st step to factor a trinomial equation? _____

2. What's the 2nd step to factor a trinomial? _____

3. How do you know it works? _____

4. What are the C term factors? $x^2 + 4x + 3$

Which ones add to the b term?   _____

____ + ____ = ____   (x + ___) (x + ___)

5. Altogether, factor the trinomial. $x^2 + 6x + 5$

(x + ___) (x + ___)

#2 1. What if C term has more factors? _____

2. What are the C term factors? $x^2 + 7x + 12$

Which ones add to the b term?   _____

____ + ____ = ____   (x + ___) (x + ___)

3. What are the C term factors? $x^2 + 9x + 14$

Which ones add to the b term?   _____

____ + ____ = ____   (x + ___) (x + ___)

4. Altogether, factor the trinomial. $x^2 + 7x + 10$

(x + ___) (x + ___)

#3 Factor each as 1 or more than 1.

1. $y = x^2 + 2x + 1$ $(\_\_ + \_\_)(\_\_ + \_\_)$ $y = x^2 + 18x + 17$ $(\_\_ + \_\_)(\_\_ + \_\_)$ Calculator? yes no

2. $y = x^2 + 20x + 19$ $(\_\_ + \_\_)(\_\_ + \_\_)$ $y = x^2 + 8x + 7$ $(\_\_ + \_\_)(\_\_ + \_\_)$

3. $y = x^2 + 6x + 5$ $(\_\_ + \_\_)(\_\_ + \_\_)$ $y = x^2 + 14x + 13$ $(\_\_ + \_\_)(\_\_ + \_\_)$

4. $y = x^2 + 7x + 10$ $(\_\_ + \_\_)(\_\_ + \_\_)$ $y = x^2 + 5x + 4$ $(\_\_ + \_\_)(\_\_ + \_\_)$

5. $y = x^2 + 7x + 12$ $(\_\_ + \_\_)(\_\_ + \_\_)$ $y = x^2 + 7x + 6$ $(\_\_ + \_\_)(\_\_ + \_\_)$

6. $y = x^2 + 6x + 8$ $(\_\_ + \_\_)(\_\_ + \_\_)$ $y = x^2 + 4x + 4$ $(\_\_ + \_\_)(\_\_ + \_\_)$

Review 1. $x^2 + 6x + 5$ What's the 1st step to factor a trinomial equation? _____   Calculator? yes no

2. What's the 2nd step to factor a trinomial? _____

3. How do you know it works? _____

4. What if C term has more factors? _____

5. Where do the numbers go? _____

6. How do you know it works? _____

7. $y = x^2 + 9x + 8$ $(\_\_ + \_\_)(\_\_ + \_\_)$ $y = x^2 + 5x + 6$ $(\_\_ + \_\_)(\_\_ + \_\_)$

8. $y = x^2 + 11x + 10$ $(\_\_ + \_\_)(\_\_ + \_\_)$ $y = x^2 + 13x + 12$ $(\_\_ + \_\_)(\_\_ + \_\_)$

Ch 4 Ls 3 Factor with A terms and more factors. 161

_____ #1 #2 ____/ 11 #3 ____/ 8 R ____/ 11 T ____/ 30 _____
         Name   Checker

#1 1. $2x^2 + 8x + 3$ How do you factor this A term? _____
    2. Name 2 ways #1's equation can factor. (2x   )(x   )   (2x   )(x   )
    3. How do you know which works? _____

        4. What are the A term factors? $3x^2 + 7x + 2$

            Which factors add to 7x?   (3x   ?) (x   ?)

            ____ + ____ = 7x   (3x ____) (x ____)

        5. What are the A term factors? $5x^2 + 8x + 3$

            Which factors add to 8x?   (5x   ?) (x   ?)

            ____ + ____ = 8x   (5x ____) (x ____)

#2 1. What's the 1st step to the factor shortcut? _____
    2. What happens after A term factors? _____
    3. What do you look for in C term factors? _____
    4. What does the 2nd fact do? _____

        5. What's the main fact? $3x^2 + 13x + 4$

            What's the 2nd fact?   (3x   ?) (x   ?)

                                           (3x ____) (x ____)

        6. What's the main fact? $2x^2 + 8x + 6$

            What's the 2nd fact?   (2x   ?) (x   ?)

                                           (2x ____) (x ____)

#3 Factor the final facts and C facts.   Calculator? yes no

1. $2x^2 + 7x + 3$ $(2x \;\underline{})(x \;\underline{})$ $\underline{} + \underline{} = 7x$

2. $3x^2 + 10x + 3$ $(3x \;\underline{})(x \;\underline{})$ $\underline{} + \underline{} = 10x$

3. $7x^2 + 11x + 3$ $(7x \;\underline{})(x \;\underline{})$ $\underline{} + \underline{} = 11x$

4. $2x^2 + 11x + 5$ $(2x \;\underline{})(x \;\underline{})$ $\underline{} + \underline{} = 11x$

5. $5x^2 + 8x + 3$ $(5x \;\underline{})(x \;\underline{})$ $\underline{} + \underline{} = 8x$

6. $2x^2 + 7x + 5$ $(2x \;\underline{})(x \;\underline{})$ $\underline{} + \underline{} = 7x$

7. $5x^2 + 16x + 3$ $(5x \;\underline{})(x \;\underline{})$ $\underline{} + \underline{} = 16x$

8. $2x^2 + 5x + 3$ $(2x \;\underline{})(x \;\underline{})$ $\underline{} + \underline{} = 5x$

Review 1. $2x^2 + 6x + 3$ How do you factor this A term? _____ Calculator? yes no

2. Name 2 ways #1's equation can factor. $(2x \quad)(x \quad)$ $(2x \quad)(x \quad)$

3. How do you know which works? _____

4. What's the 1st step to the factor shortcut? _____

5. What happens after A term factors? _____

6. What do you look for in C term factors? _____

7. What does the 2nd fact do? _____

Factor thinomials.

8. $4x^2 + 10x + 6$ $(\;x \;\underline{})(\;x \;\underline{})$ $\underline{} + \underline{} = 10x$

9. $8x^2 + 12x + 4$ $(\;x \;\underline{})(\;x \;\underline{})$ $\underline{} + \underline{} = 12x$

10. $6x^2 + 14x + 8$ $(\;x \;\underline{})(\;x \;\underline{})$ $\underline{} + \underline{} = 14x$

11. $8x^2 + 12x + 4$ $(\;x \;\underline{})(\;x \;\underline{})$ $\underline{} + \underline{} = 12x$

Review Problems 163

_____ #1 #2 #3 ____ / 24   #4 #5 ____ / 16   T ____ / 40
Name

#1 1. Common Factor _____

2. Divide Binomials _____

3. Factor Trinomials _____

#2 1. $8a + 4 =$ _____ $2b - 6 =$ _____ $9c^2 + 3c =$ _____   Calculator?
Simplify   yes   no
these.   2. $10d - 12 =$ _____ $4b^2 - 3b =$ _____ $10c^2 + 5c =$ _____

3. $5g - 2.5 =$ _____ $6x^2 - 4x =$ _____ $8y^2 - 6y =$ _____

4. $\dfrac{6x^2 + 2}{8x^2 + 4}$ $\dfrac{2a - 4}{3a - 6}$ $\dfrac{4b^2 - 6b}{6}$

   _____ = ___   _____ = ___   _____ = ___

5. $\dfrac{2c^2 + 4c}{4c}$ $\dfrac{6d^2 - 4d}{2d}$ $\dfrac{4x}{8x - 4}$

   _____ = ___   _____ = ___   _____ = ___

#3 Factor these quadratics.   Calculator?
                                                                                          yes   no

1. $y = x^2 + 4x + 3$ $y = x^2 + 18x + 17$

   (___ + ___)(___ + ___)   (___ + ___)(___ + ___)

2. $y = x^2 + 16x + 15$ $y = x^2 + 9x + 8$

   (___ + ___)(___ + ___)   (___ + ___)(___ + ___)

3. $y = x^2 + 12x + 11$ $y = x^2 + 20x + 19$

   (___ + ___)(___ + ___)   (___ + ___)(___ + ___)

#4 Continued from Part 3.   Calculator? yes no

4. $y = x^2 + 7x + 12$ $y = x^2 + 5x + 6$

(___ + ___)(___ + ___)   (___ + ___)(___ + ___)

5. $y = x^2 + 8x + 15$ $y = x^2 + 6x + 9$

(___ + ___)(___ + ___)   (___ + ___)(___ + ___)

6. $y = x^2 + 9x + 14$ $y = x^2 + 8x + 16$

(___ + ___)(___ + ___)   (___ + ___)(___ + ___)

#5 Factor to find the binomials. Does it equal the C facts?   Calculator? yes no

1. $4x^2 + 13x + 3$ (___ + ___)(___ + ___)   ___ + ___ = 13x

2. $5x^2 + 9x + 4$ (___ + ___)(___ + ___)   ___ + ___ = 9x

3. $8x^2 + 11x + 3$ (___ + ___)(___ + ___)   ___ + ___ = 11x

4. $3x^2 + 16x + 5$ (___ + ___)(___ + ___)   ___ + ___ = 16x

5. $9x^2 + 12x + 3$ (___ + ___)(___ + ___)   ___ + ___ = 12x

6. $4x^2 + 12x + 5$ (___ + ___)(___ + ___)   ___ + ___ = 12x

7. $6x^2 + 11x + 3$ (___ + ___)(___ + ___)   ___ + ___ = 11x

8. $3x^2 + 7x + 4$ (___ + ___)(___ + ___)   ___ + ___ = 7x

9. $5x^2 + 11x + 2$ (___ + ___)(___ + ___)   ___ + ___ = 11x

10. $8x^2 + 10x + 3$ (___ + ___)(___ + ___)   ___ + ___ = 10x

Ch 5 Ls 1 Factor trinomials with negatives.   165

_____ #1 #2 ____/ 9 #3 ____/ 12 R ___/ 10 T ____/ 31 _____
             Name   Checker

#1 1. $x^2 - 2x - 3$ Negative C term. How many sets are there? _____
    2. What is the shortcut to know which ones work? _____
    3. 1 x - 3 or - 1 x 3 Which factors does it use? _____
    4. B term is negative. C is positive. What do you know? _____

#2 1. What are the C factors? $x^2 - 4x - 5$
        1 x - 5   - 1 x 5   (x   ?) (x   ?)   **What are the binomials?**

                                  (x ____) (x ____)

    2. What are the C factors? $x^2 + 5x - 4$
        1 x - 4   - 1 x 4   2 x - 2   (x   ?) (x   ?)   Find the binomials.

                                  (x ____) (x ____)

    3. What are the C factors? $x^2 - 5x + 4$
        - 1 x - 4   - 2 x - 2   (x   ?) (x   ?)   Find the binomials.

                                  (x ____) (x ____)

    4. What are the C factors? $x^2 - 9x + 8$
        - 2 x - 4   - 1 x - 8   (x   ?) (x   ?)   **What are the binomials?**

                                  (x ____) (x ____)

    5. What are the C factors? $x^2 - 8x + 12$
        - 1 x - 12   - 2 x - 6   - 3 x - 4   (x   ?) (x   ?)   **What are the binomials?**

                                  (x ____) (x ____)

#3 Find the C factors. Which ones work?   Calculator? yes no

1. $x^2 - 5x - 6$ $x^2 + x - 6$ $x^2 + 5x - 6$

_____   _____   _____

(__ + __)(__ + __)   (__ + __)(__ + __)   (__ + __)(__ + __)

2. $x^2 + 7x - 8$ $x^2 - 2x - 8$ $x^2 - 7x - 8$

_____   _____   _____

(__ + __)(__ + __)   (__ + __)(__ + __)   (__ + __)(__ + __)

3. $x^2 - 13x - 14$ $x^2 - 5x - 14$ $x^2 + 5x - 14$

_____   _____   _____

(__ + __)(__ + __)   (__ + __)(__ + __)   (__ + __)(__ + __)

4. $x^2 + 3x - 10$ $x^2 + 9x - 10$ $x^2 - 3x - 10$

_____   _____   _____

(__ + __)(__ + __)   (__ + __)(__ + __)   (__ + __)(__ + __)

Review 1. $x^2 - 2x - 3$ Negative C term. How many sets are there? _____   Calculator?
2. What is the shortcut to know which ones work? _____   yes no
3. 1 x -3 or -1 x 3 Which factors does it use? _____
4. B term is negative. C is positive. What do you know? _____

5. $x^2 + 3x - 4$ $x^2 - 3x - 4$ $x^2 - 5x - 6$

(__ + __)(__ + __)   (__ + __)(__ + __)   (__ + __)(__ + __)

6. $x^2 - 2x - 8$ $x^2 + 7x - 8$ $x^2 - 6x - 8$

(__ + __)(__ + __)   (__ + __)(__ + __)   (__ + __)(__ + __)

Ch 5 Ls 2   Quadratics with no B term.   167

_____ #1 #2 ____ / 9 #3 ____ / 6 R ____ / 11 T ____ / 26 _____
         Name   Checker

#1 1. What's the 1st step to factor a A negative? _____
_____

   2. What happens to the signs? _____

      3. What's the 1st step with negative A? $-x^2 + 6x - 3$

         How do the signs change?   _____

                                          _____

      4. What's the 1st step with negative A? $-x^2 - 4x - 2$

         How do the signs change?   _____

                                          _____

#2 1. What changes a fraction in an equation? _____
   2. What changes a decimal in an equation? _____
   3. What's the problem with the equation made? _____

      4. How do you get rid of the fraction? $x^2 + \frac{1}{2}x + 3$

         Multiply _____ __(   )   Finish it.

                                              _____

      5. How do you get rid of the fraction? $2x^2 - 0.2x + 0.05$

         Multiply _____ __(   )   Finish it.

                                              _____

#3 Change to a different 1st sign, then factor it. (or circle Not.)   Calculator?
                                                                        yes   no

1. $-3x^2 + 2x - 5$ $-2x^2 - 2x + 6$

   $-(\_\_\_\ \_\_\_\ \_\_\_)$ $-(\_\_\_\ \_\_\_\ \_\_\_)$

   $(\_\_\ \_\_)(\_\_\ \_\_)$ Not $(\_\_\ \_\_)(\_\_\ \_\_)$ Not

2. $-x^2 - 4x + 5$ $-4x^2 + 8x - 5$

   $-(\_\_\_\ \_\_\_\ \_\_\_)$ $-(\_\_\_\ \_\_\_\ \_\_\_)$

   $(\_\_\ \_\_)(\_\_\ \_\_)$ Not $(\_\_\ \_\_)(\_\_\ \_\_)$ Not

3. $-3x^2 + 8x - 6$ $-6x^2 - 3x + 8$

   $-(\_\_\_\ \_\_\_\ \_\_\_)$ $-(\_\_\_\ \_\_\_\ \_\_\_)$

   $(\_\_\ \_\_)(\_\_\ \_\_)$ Not $(\_\_\ \_\_)(\_\_\ \_\_)$ Not

Review 1. Name 3 steps to factor negatives. _____   Calculator?
        _____   yes   no

2. What happens to the signs? _____

3. What changes a fraction in an equation? _____

4. What changes a decimal in an equation? _____

5. What's the problem with the equation made? _____

6. $2x^2 + \frac{1}{2}x + 4$ $3x^2 + \frac{1}{2}x - 3$

   $(\_\_\_\ \_\_\_\ \_\_\_)$ $(\_\_\_\ \_\_\_\ \_\_\_)$

7. $3x^2 + \frac{1}{3}x + 4$ $x^2 + \frac{1}{4}x - 2$

   $(\_\_\_\ \_\_\_\ \_\_\_)$ $(\_\_\_\ \_\_\_\ \_\_\_)$

8. $4x^2 + \frac{1}{2}x - 2$ $7x^2 + \frac{1}{3}x + 1$

   $(\_\_\_\ \_\_\_\ \_\_\_)$ $(\_\_\_\ \_\_\_\ \_\_\_)$

Ch 5 Ls 3 Factor trinomials with negatives. 169

_____ #1 #2 ____ / 11 #3 ____ / 8 R ____ / 10 T ____ / 29 _____
           Name   Checker

#1 1. Name 3 steps to factor negatives. _____

_____

    2. How do you know it works? _____

        3. What's the main fact? $2x^2 - 9x - 5$

            What's the 2nd fact?   (2x   ?) (x   ___)

                                                 (2x   ___) (x   ___)

        4. What's the main fact? $3x^2 + 20x - 7$

            What's the 2nd fact?   (3x   ?) (x   ___)

                                                 (3x   ___) (x   ___)

        5. What's the main fact? $2x^2 - 9x + 4$

            What's the 2nd fact?   (2x   ?) (x   ___)

                                                 (2x   ___) (x   ___)

#2 1. $x^2 - 9$ What does A term have to be to make B term disappear? _____

      2. What numbers do C term have to be? _____

      3. Why does C term have to be negative? _____

      4. Name 3 steps to no B term? _____

_____

        5. How do you get rid of a B term?   (a + 3)(a - 3)

                                                      _____

        6. How do you get rid of a B term?   (x + 4)(x - 4)

                                                      _____

#3 What's the correct answer?   Calculator? yes no

1. $3x^2 - 2x - 5$ (3x ___)(x ___)   ___ + ___ = -2x

2. $3x^2 + 2x - 5$ (3x ___)(x ___)   ___ + ___ = +2x

3. $2x^2 + x - 3$ (2x ___)(x ___)   ___ + ___ = +1x

4. $2x^2 - 5x - 3$ (2x ___)(x ___)   ___ + ___ = -5x

5. $5x^2 - 14x - 3$ (5x ___)(x ___)   ___ + ___ = -14x

6. $5x^2 + 2x - 3$ (5x ___)(x ___)   ___ + ___ = +2x

7. $7x^2 - 2x - 5$ (7x ___)(x ___)   ___ + ___ = -2x

8. $7x^2 + 2x - 5$ (7x ___)(x ___)   ___ + ___ = +2x

Review 1. Name 3 steps to factor negatives. _____   Calculator? yes no
_____

2. How do you solve a negative A term? _____
3. $x^2 - 9$ What does A term have to be to make B term disappear? _____
4. What numbers do C term have to be? _____
5. Why does C term have to be negative? _____
6. Name 3 steps to no B term? _____
_____

7. $4x^2 - 10x - 6$ (2x ___)(2x ___) ___ + ___ = -10x

8. $8x^2 - 4x - 4$ (4x ___)(2x ___) ___ + ___ = -4x

9. $6x^2 + 2x - 4$ (3x ___)(2x ___) ___ + ___ = +2x

10. $4x^2 - 10x - 6$ (4x ___)(x ___)   ___ + ___ = -10x

Ch 5 Ls 4 How to find x intercepts. 171

_____ #1 #2 ____/ 6 #3 ____/ 6 R ____/ 3 T ____/ 15 _____
          Name   Checker

#1 1. How do quadratics use X intercepts? _____

2. If you can't factor, what do you know? _____

3. Name 3 steps to FOX. _____

_____

#2 1. What's the 1st step to find X intercepts? $y = x^2 - 4x + 5$

After you factor it, what's the next step? $y = (\_\_\ \_\_)(\_\_\ \_\_)$

Set it to 0. What happens next? $\_\_ = (_____)(_____)$

What are the X intercepts? $\_\_ = _____$ $\_\_ = _____$

$\_\_ = \_\_$ $\_\_ = \_\_$

2. What's the 1st step to find X intercepts? $y = 2x^2 - 9x - 5$

After you factor it, what's the next step? $y = (\_\_\ \_\_)(\_\_\ \_\_)$

Set it to 0. What happens next? $\_\_ = (_____)(_____)$

What are the X intercepts? $\_\_ = _____$ $\_\_ = _____$

$\_\_ = \_\_$ $\_\_ = \_\_$

3. Mentally find the x intercepts. $y = x^2 - 5x + 6$

After you factor it, what's the next step? $y = (\_\_\ \_\_)(\_\_\ \_\_)$

What are the X intercepts? $\_\_ = (_____)(_____)$

$\_\_ = \_\_$ $\_\_ = \_\_$

#3 Factor each of these equations. Find X interecepts.   Calculator? yes no

1. $y = x^2 - x - 2$ $y = x^2 + 5x - 6$

   y = (___ ___)(___ ___)   y = (___ ___)(___ ___)

   ___ = ___   ___ = ___   ___ = ___   ___ = ___

   ___ = ___   ___ = ___   ___ = ___   ___ = ___

   X intercepts are ___, ___ and ___, ___.   X intercepts are ___, ___ and ___, ___.

2. $y = x^2 + 5x + 4$ $y = x^2 - 2x - 8$

   y = (___ ___)(___ ___)   y = (___ ___)(___ ___)

   ___ = ___   ___ = ___   ___ = ___   ___ = ___

   ___ = ___   ___ = ___   ___ = ___   ___ = ___

   X intercepts are ___, ___ and ___, ___.   X intercepts are ___, ___ and ___, ___.

3. $y = x^2 + 5x + 6$ $y = x^2 + 7x + 10$

   y = (___ ___)(___ ___)   y = (___ ___)(___ ___)

   ___ = ___   ___ = ___   ___ = ___   ___ = ___

   ___ = ___   ___ = ___   ___ = ___   ___ = ___

   X intercepts are ___, ___ and ___, ___.   X intercepts are ___, ___ and ___, ___.

Review 1. How do quadratics use X intercepts? _____

2. If you can't factor, what do you know? _____

3. Name 3 steps to FOX. _____

_____

Ch 5 Ls 5 How to graph a trinomial. 173

_____ #1 #2 ____ / 6 #3 ____ / 4 R ____ / 4 T ____ / 14 _____
       Name   Checker

#1 1. What is SOY? _____
       2. What does AVX stand for? _____
       3. What's the last step to make a parabola? _____
       4. What are 3 steps to graph a parabola? _____

#2 1. What's first to graph an equation? $y = x^2 + 4x + 3$

       Where is the axis and vertex? Slope: ____ - o + Opens: ____ Y int: ____

       Find the X intercepts.   _____

       What does the graph look like? _____

```
         8
         6
         4
         2
-8 -6 -4 -2  0  2  4  6  8
        -2
        -4
        -6
        -8
```

       2. What's first to graph an equation? $y = x^2 + 6x + 5$

       Where is the axis and vertex? Slope: ____ - o + Opens: ____ Y int: ____

       Find the X intercepts.   _____

       What does the graph look like? _____

```
         8
         6
         4
         2
-8 -6 -4 -2  0  2  4  6  8
        -2
        -4
        -6
        -8
```

#3 Solve the equations for each part to make a graph.   Calculator? yes   no

1. $y = 3x^2 - 7x - 6$

Slope: - o +   Opens: ____   Y int: ____
Axis ____   Vertex ____   X int ____ ____

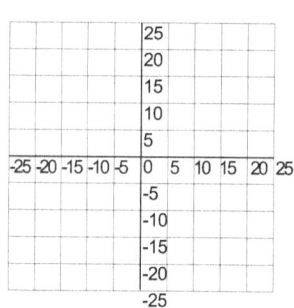

$y = 2x^2 + 4x + 6$

Slope: - o +   Opens: ____   Y int: ____
Axis ____   Vertex ____   X int ____ ____

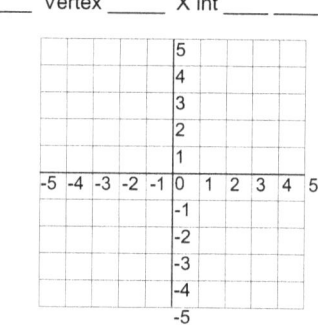

2. $y = -x^2 + 2x - 1$

Slope: - o +   Opens: ____   Y int: ____
Axis ____   Vertex ____   X int ____ ____

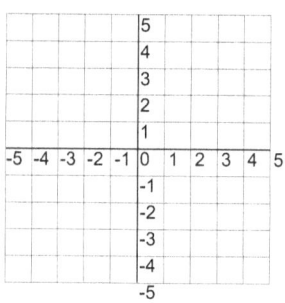

$y = \frac{1}{2}x^2 + 2x + 4$

Slope: - o +   Opens: ____   Y int: ____
Axis ____   Vertex ____   X int ____ ____

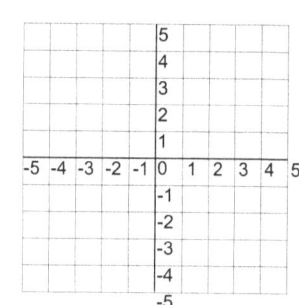

Review

1. What is SOY and what does it do? _____
2. What does AVX stand for? _____
3. What's the last step to make a parabola? _____
4. What are 3 steps to graph a parabola? _____

Review Problems 175

_____   #1 to #4 ____ / 19   #5 #6 ____ / 6 T ___ / 25
            Name

#1 1. Factor Negatives _____
    2. Shortcut _____
    3. Equal Equations _____
    4. X Intercepts _____
    5. SOY AVX 1 _____

#2 Find the C factors. Which ones work?   Calculator?
                                                             yes   no

1. $x^2 + 3x - 4$ $x^2 + 2x - 8$ $x^2 - 3x - 10$

    (__ + __)(__ + __)   (__ + __)(__ + __)   (__ + __)(__ + __)

2. $x^2 - 2x - 24$ $x^2 + 5x - 150$ $x^2 - 6x - 16$

    (__ + __)(__ + __)   (__ + __)(__ + __)   (__ + __)(__ + __)

#3 What's the correct answer?   Calculator?
                                                             yes   no

1. $3x^2 - 3x - 6$ (__ x __)(__ x __)   ___ + ___ = -3x

2. $5x^2 + 8x - 4$ (__ x __)(__ x __)   ___ + ___ = +8x

3. $6x^2 + 2x - 4$ (__ x __)(__ x __)   ___ + ___ = +2x

4. $4x^2 - 2x - 6$ (__ x __)(__ x __)   ___ + ___ = -2x

#4 Change to a different 1st sign, then factor it.   Calculator?
                                                             yes   no

1. $-3x^2 + 8x - 5$ -(___ ___ ___) is _____

2. $-4x^2 + 8x - 5$ -(___ ___ ___) is _____

3. $-x^2 + 4x + 12$ -(___ ___ ___) is _____

4. $-2x^2 - x + 6$ -(___ ___ ___) is _____

#5 Find the x intercept for these equations.   Calculator? yes no

1. $y = x^2 - 4x - 12$ $y = x^2 + 2x - 8$

   y = (___ ___)(___ ___)   y = (___ ___)(___ ___)

   ___ = _____   ___ = _____   ___ = _____   ___ = _____

   ___ = ___   ___ = ___   ___ = ___   ___ = ___

   X intercepts are ___, ___ and ___, ___.   X intercepts are ___, ___ and ___, ___.

2. $y = x^2 + 2x - 15$ $y = x^2 - 3x - 18$

   y = (___ ___)(___ ___)   y = (___ ___)(___ ___)

   ___ = _____   ___ = _____   ___ = _____   ___ = _____

   ___ = ___   ___ = ___   ___ = ___   ___ = ___

   X intercepts are ___, ___ and ___, ___.   X intercepts are ___, ___ and ___, ___.

#6 Solve the equations for each part to make a graph.   Calculator? yes no

1. $y = 2x^2 + 4x - 6$ $y = 4x^2 + 2x - 2$

Slope: - o +   Opens: ____ Y int: ____   Slope: - o +   Opens: ____ Y int: ____

Vertex ____ X int ____ and ____   Vertex ____ X int ____ and ____

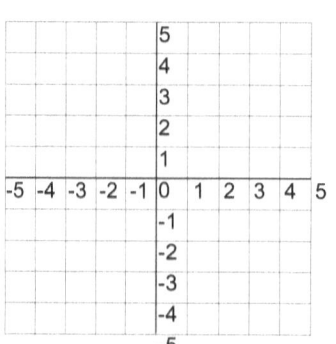
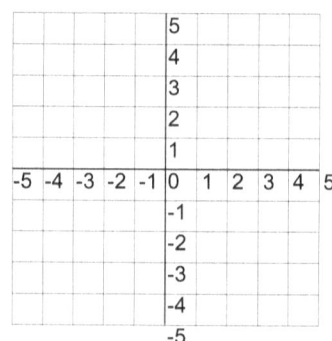

Ch 6 Ls 1 Complete the Square. 177

_____ #1 #2 ____/ 10 #3 ____/ 6 R ____/ 6 T ____/ 22 _____
            Name   Checker

#1   1. What does Complete the Square do? _____

     2. When can an equation use it? _____

     3. What are 2 steps to a new C term? _____

     4. What happens to the new C term? _____

     5. What's the 3rd step? _____

     6. What's the 4th step? _____

#2   1. It's standard. What's the 1st step? $x^2 + 4x - 1 = 0$

         Use 2 steps to find a new C term.   _____

                   Change the equation.   Divide: ____ Square: ____

                                                   _____

     2. It's standard. What's the 1st step? $x^2 + 2x + 5 = 0$

         Use 2 steps to find a new C term.   _____

                   Change the equation.   Divide: ____ Square: ____

                                                   _____

     3. Use 2 steps to find a new C term. $x^2 + 8x = 0$

                   Change the equation.   Divide: ____ Square: ____

                                                   _____

     4. Use 2 steps to find a new C term. $x^2 + 10x = 0$

                   Change the equation.   Divide: ____ Square: ____

                                                   _____

#3 Use these equations to create another equation.   Calculator? yes   no

1. $x^2 + 2x + 5 = 0$ $x^2 + 6x - 2 = 0$

   _____   _____

   Divide: ___ Square: ___   Divide: ___ Square: ___
   New Eq _____   New Eq _____
   Factor _____   Factor _____

2. $x^2 + 4x - 7 = 0$ $x^2 + 4x + 1 = 0$

   _____   _____

   Divide: ___ Square: ___   Divide: ___ Square: ___
   New Eq _____   New Eq _____
   Factor _____   Factor _____

3. $x^2 + 10x + 4 = 0$ $x^2 + 2x - 6 = 0$

   _____   _____

   Divide: ___ Square: ___   Divide: ___ Square: ___
   New Eq _____   New Eq _____
   Factor _____   Factor _____

Review 1. What does Complete the Square do? _____

2. When can an equation use it? _____

3. What are 2 steps to a new C term? _____

4. What happens to the new C term? _____

5. What's the 3rd step? _____

6. What's the 4th step? _____

Ch 6 Ls 2 Complete the square review. 179

_____ #1 #2 ____ /9 #3 ____ /4 R ____ /9 T ____ /22 _____
         Name   Checker

#1 1. Name 3 things an equation has to complete the square. _____

_____

2. What does CB22 mean? _____
3. What does BS tell you about the new C? _____
4. How does it always factor? _____
5. What does SS2 in FSS2 show? _____

      6. What happens to the - 1? $x^2 + x - 1 = 0$

          2 steps to find a new C.   _____

            What happens to it?   _____

          What does F in FSS2 do?   _____

         What is the positive answer?   _____

           Find the negative answer.   _____

                                        _____

#2 1. What happens when the 1st term starts with a decimal? _____

_____

    2. What does this equation multiply by? $0.5 x^2 + 2x = 1$

        Multiply by _____. What does it equal?   _____

                                        _____

    3. What does this equation multiply by? $0.02 x^2 + x = 1$

                                        _____

#3 1. $x^2 + 1x = 4$ $x^2 + 3x = 5$ Calculator?
Use Complete   yes   no
the Square. 1st step? _____   _____

             New C?   _____   _____

                        _____   _____
             Positive Ans.
                        _____   _____
             Negative Ans.
                        _____   _____

         2. $x^2 + \frac{1}{2}x = 2$ $x^2 + \frac{1}{4}x = -3$

             1st step? _____   _____

             New C?   _____   _____

                        _____   _____
             Positive Ans.
                        _____   _____
             Negative Ans.
                        _____   _____

Review 1. Name 3 things an equation has. _____   Calculator?
        _____   yes   no

        2. What does CB22 mean? _____
        3. What does BS tell you about the new C? _____
        4. How does it always factor? _____
        5. What does SS2 in FSS2 show? _____
        6. What happens when the 1st term starts with a decimal? _____
        _____

            What do these equartions muliply by? What does it equal?
        7.
            $0.6x^2 + 2x = -4$ $0.002x^2 + 3x = 5$ $\frac{1}{7}x^2 + 3x = 6$

            Multiply _____   Multiply _____   Multiply _____

It equals _____   _____   _____

Review Problems 181

_____ #1 #2 ____ / 11 #3 ____ / 8 Total ____ / 19
             Name

#1 1. Complete the Square _____
     2. CB22 BS F 222 _____

#2 1. $x^2 + 1x + 4 = 0$ $x^2 + 3x - 5 = 0$ Calculator?
                                                                       yes   no

1st step? _____   _____

New C?   _____   _____

Positive Ans. _____   _____

Negative Ans. _____   _____

              _____   _____

2. $x^2 + \frac{1}{2}x + 2 = 0$ $x^2 + \frac{1}{4}x - 3 = 0$

1st step? _____   _____

New C?   _____   _____

Positive Ans. _____   _____

Negative Ans. _____   _____

              _____   _____

3. $x^2 + 1x + 4 = 0$ $x^2 + 3x - 5 = 0$

1st step? _____   _____

New C?   _____   _____

Positive Ans. _____   _____

Negative Ans. _____   _____

              _____   _____

#3 1. $x^2 + 0.5x + 6 = 0$ $x^2 + 0.75x - 7 = 0$ Calculator? yes no

Use Complete the Square.

1st step? _____   _____

New C? _____   _____

Positive Ans. _____   _____

Negative Ans. _____   _____

2. $x^2 + 0.6x + 3 = 0$ $x^2 + 0.8x - 8 = 0$

1st step? _____   _____

New C? _____   _____

Positive Ans. _____   _____

Negative Ans. _____   _____

3. $x^2 + 2x + 9 = 0$ $x^2 + 5x - 1 = 0$

1st step? _____   _____

New C? _____   _____

Positive Ans. _____   _____

Negative Ans. _____   _____

#4 1. What does Complete the Square do? _____

2. When can an equation use it? _____

3. What are 2 steps to a new C term? _____

4. What happens to the new C term? _____

Ch 7 Ls 1 The Quadratic Formula/Discriminant. 183

_____ #1 #2 ____/ 9 #3 ____/ 9 R ___/ 10 T ____/ 28 _____
Name   Checker

#1 1. What formula starts the quadratic formula? _____
    2. What part of quadratic formula decides X intercepts? _____
    3. What's the 1st step to find it? _____
    4. What happens after B squared? _____

#2 1. What are 2 steps to the determinents?   A is 4   B is 6   C is 3

                    What did it find?   _____ - 4( ) ( )

                    Pos Neg 1 point   _____

    2. What are 2 steps to the determinents?   A is 2   B is 5   C is 1

                    What did it find?   _____ - 4( ) ( )

                    Pos Neg 1 point   _____

    3. Find the determinents. $y = 2x^2 + 6x + 1$

                    What did it find?   _____ - 4( ) ( )

                    Pos Neg 1 point   _____

    4. Find the determinents. $y = 5x^2 + 3x + 1$

                    What did it find?   _____ - 4( ) ( )

                    Pos Neg 1 point   _____

    5. Find the determinents. $y = x^2 + 5x + 1$

                    What did it find?   _____ - 4( ) ( )

                    Pos Neg 1 point   _____

#3 Find the corresponding point. Calculator? yes no

1. $2x^2 + 3x + 5$ $2x^2 + 4x + 2$ $2x^2 + 5x + 1$
   ___ - 4( )( )   ___ - 4( )( )   ___ - 4( )( )

   ──────────   ──────────   ──────────
   Pos Neg 1 point   Pos Neg 1 point   Pos Neg 1 point

2. $3x^2 + 6x + 4$ $5x^2 + 4x + 1$ $4x^2 + 2x + 4$
   ___ - 4( )( )   ___ - 4( )( )   ___ - 4( )( )

   ──────────   ──────────   ──────────
   Pos Neg 1 point   Pos Neg 1 point   Pos Neg 1 point

3. $x^2 + 4x + 5$ $6x^2 + 6x + 7$ $2x^2 + 4x + 3$
   ___ - 4( )( )   ___ - 4( )( )   ___ - 4( )( )

   ──────────   ──────────   ──────────
   Pos Neg 1 point   Pos Neg 1 point   Pos Neg 1 point

Review 1. Where does the quadratic formula come from? _____ Calculator? yes no
2. What part of the quadratic formula predicts X intercepts? _____
3. What are 3 places Complete the Square uses. _____
4. How does the determinents show if there's X intercepts? _____

5. $x^2 + 2x + 3$ $2x^2 + 5x + 3$ $x^2 + 4x + 3$
   ___ - 4( )( )   ___ - 4( )( )   ___ - 4( )( )

   ──────────   ──────────   ──────────
   Pos Neg 1 point   Pos Neg 1 point   Pos Neg 1 point

6. $4x^2 + 3x + 2$ $3x^2 + 4x + 1$ $2x^2 + 3x + 4$
   ___ - 4( )( )   ___ - 4( )( )   ___ - 4( )( )

   ──────────   ──────────   ──────────
   Pos Neg 1 point   Pos Neg 1 point   Pos Neg 1 point

Ch 7 Ls 2 Solve the Quadratic Formula.   185

_____ #1 #2 ____/8 #3 ____/6 R ___/7 T ____/21 _____
Name   Checker

#1 1. What does the quadratic formula do? _____
2. What formula starts the quadratic formula? _____
3. What's the 1st step? _____
4. What 3 steps solve the equation? _____

#2 1. a is 2   b is 5   c is 2
What's the positive answer? $\dfrac{-5 \pm \sqrt{16}}{2(1)}$

Find the negative answer.

---

2. a is 0.5   b is 5   c is -1
What's the positive answer? $\dfrac{-3 \pm \sqrt{27}}{2(0.5)}$ 5.2

Find the negative answer.

---

3. a is -2   b is -3   c is 1
What's the positive answer? $\dfrac{-2 \pm \sqrt{17}}{2(-2)}$ 4.1

Find the negative answer.

---

4. a is 1   b is 2   c is -1
What's the positive answer? $\dfrac{-2 \pm \sqrt{8}}{2(1)}$ 2.8

Find the negative answer.

#3 **1.** $\dfrac{-2 \pm \sqrt{1}}{2(1)}$ $\dfrac{-6 \pm \sqrt{100}}{2(2)}$ Calculator? yes no

Solve these Quadratic Equations.

_____   _____

_____   _____

2. $\dfrac{-5 \pm \sqrt{81}}{2(1)}$ $\dfrac{-5 \pm \sqrt{36}}{2(1)}$

_____   _____

_____   _____

_____   _____

Review 1. What does the quadratic formula do? _____   Calculator?

2. What formula starts the quadratic formula? _____   yes no

3. How does it use the Determinent? _____

4. What 3 steps solve the equation? _____

If you need more space use another paper or on a side.

5. $y = 2x^2 + 3x - 5$ positive $-\dfrac{\ \ +\sqrt{\ \ \ }}{2(\ \ )} = $ _____

_____ $- 4(\ \ )(\ \ )$

_____   negative $-\dfrac{\ \ -\sqrt{\ \ \ }}{2(\ \ )} = $ _____

6. $y = 3x^2 + 5x + 1$ positive $-\dfrac{\ \ +\sqrt{\ \ \ }}{2(\ \ )} = $ _____

_____ $- 4(\ \ )(\ \ )$

_____   negative $-\dfrac{\ \ -\sqrt{\ \ \ }}{2(\ \ )} = $ _____

7. $y = 4x^2 - 3x - 2$ positive $-\dfrac{\ \ +\sqrt{\ \ \ }}{2(\ \ )} = $ _____

_____ $- 4(\ \ )(\ \ )$

_____   negative $-\dfrac{\ \ -\sqrt{\ \ \ }}{2(\ \ )} = $ _____

Ch 7 Ls 3 More discriminant problems.   187

_____ #1 #2 ____ / 8   #3 ____ / 9 R ____ / 9 T ____ / 26 _____
Name   Checker

#1 1. How does B sign change if there are X intercepts? _____
   2. If there's 1 negative for A or C, does it have x intercepts? _____
   3. If both A and C have negatives, does it have x intercepts? _____

#2 1. What are 2 steps to the discriminent? A is 1   B is - 6   C is - 3

   What did it find?   _____ - 4( ) ( )

   Pos Neg 1 point   _____

   2. What are 2 steps to the discriminent? A is - 2   B is 5   C is - 4

   What did it find?   _____ - 4( ) ( )

   Pos Neg 1 point   _____

   3. Find the discrirminent. $y = -0.5x^2 + 5x - 2$

   What did it find?   _____ - 4( ) ( )

   Pos Neg 1 point   _____

   4. Find the discrirminent. $y = 2x^2 + x - 0.1$

   What did it find?   _____ - 4( ) ( )

   Pos Neg 1 point   _____

   5. Are there X intercepts? $y = -9x^2 + 8x + 5$

   What did it find?   _____ - 4( ) ( )

   Pos Neg 1 point   _____

#3 Is it positive, negative, or 1 point?   Calculator? yes no

1. $-x^2 + 3x - 5$ $-2x^2 - 4x + 3$ $2x^2 - 3x - 4$

   ___ $-4(\ )(\ )$ ___ $-4(\ )(\ )$ ___ $-4(\ )(\ )$

   _____   _____   _____
   Pos Neg 1 point   Pos Neg 1 point   Pos Neg 1 point

2. $2x^2 - 7x + 8$ $-4x^2 + 5x - 6$ $-3x^2 - 3x + 5$

   ___ $-4(\ )(\ )$ ___ $-4(\ )(\ )$ ___ $-4(\ )(\ )$

   _____   _____   _____
   Pos Neg 1 point   Pos Neg 1 point   Pos Neg 1 point

3. $2x^2 + 4x + 5$ $-6x^2 - 3x + 7$ $2x^2 - x - 3$

   ___ $-4(\ )(\ )$ ___ $-4(\ )(\ )$ ___ $-4(\ )(\ )$

   _____   _____   _____
   Pos Neg 1 point   Pos Neg 1 point   Pos Neg 1 point

Review 1. How does B sign change if there are X intercepts? _____   Calculator?
2. If there's 1 negative for A or C, does it have x intercepts? _____   yes no
3. If both A and C have negatives, does it have x intercepts? _____

4. $x^2 + 4x - 6$ $-2x^2 - 4x - 6$ $3x^2 - 7x + 8$

   ___ $-4(\ )(\ )$ ___ $-4(\ )(\ )$ ___ $-4(\ )(\ )$

   _____   _____   _____
   Pos Neg 1 point   Pos Neg 1 point   Pos Neg 1 point

5. $5x^2 + 2x - 1$ $-3x^2 - 4x + 5$ $6x^2 - 4x - 3$

   ___ $-4(\ )(\ )$ ___ $-4(\ )(\ )$ ___ $-4(\ )(\ )$

   _____   _____   _____
   Pos Neg 1 point   Pos Neg 1 point   Pos Neg 1 point

Review Problems 189

_____ #1 to #3 _____ / 14 #4 #5 _____ / 15 T ___ / 29
      Name

#1 1. Determinent _____
    2. Quadratic Formula _____

#2 Find the corresponding point.   Calculator?
                                                            yes   no

1. $2x^2 + 3x + 5$ $2x^2 + 4x + 2$ $2x^2 + 5x + 1$

    ____ $- 4($ $)($ $)$ ____ $- 4($ $)($ $)$ ____ $- 4($ $)($ $)$

    _____   _____   _____
    Pos Neg 1 point   Pos Neg 1 point   Pos Neg 1 point

2. $3x^2 + 6x + 4$ $5x^2 + 4x + 1$ $4x^2 + 2x + 4$

    ____ $- 4($ $)($ $)$ ____ $- 4($ $)($ $)$ ____ $- 4($ $)($ $)$

    _____   _____   _____
    Pos Neg 1 point   Pos Neg 1 point   Pos Neg 1 point

#3 Solve these Quadratic Equations.   Calculator?
                                                            yes   no

1. $\dfrac{3 \pm \sqrt{49}}{2(4)}$ $\dfrac{6 \pm \sqrt{144}}{2(5)}$ $\dfrac{7 \pm \sqrt{81}}{2(3)}$

    _____   _____   _____
    _____   _____   _____
    _____   _____   _____

2. $\dfrac{-4 \pm \sqrt{16}}{2(2)}$ $\dfrac{-8 \pm \sqrt{121}}{2(1)}$ $\dfrac{9 \pm \sqrt{64}}{2(4)}$

    _____   _____   _____
    _____   _____   _____
    _____   _____   _____

#3 Is it positive, negative, or 1 point?   Calculator? yes no

1. $2x^2 + 3x + 5$ $2x^2 + 4x + 2$ $2x^2 + 5x + 1$
    ___ - 4( )( )   ___ - 4( )( )   ___ - 4( )( )
    ─────────────   ─────────────   ─────────────
    Pos Neg 1 point   Pos Neg 1 point   Pos Neg 1 point

2. $3x^2 + 6x + 4$ $5x^2 + 4x + 1$ $4x^2 + 2x + 4$
    ___ - 4( )( )   ___ - 4( )( )   ___ - 4( )( )
    ─────────────   ─────────────   ─────────────
    Pos Neg 1 point   Pos Neg 1 point   Pos Neg 1 point

3. $-x^2 + 3x - 5$ $-2x^2 - 4x + 3$ $2x^2 - 3x - 4$
    ___ - 4( )( )   ___ - 4( )( )   ___ - 4( )( )
    ─────────────   ─────────────   ─────────────
    Pos Neg 1 point   Pos Neg 1 point   Pos Neg 1 point

#4 Use Quadratic Equations.

1. $\dfrac{-4 \pm \sqrt{0.16}}{2(3)}$ $\dfrac{-6 \pm \sqrt{0.25}}{2(2)}$ $\dfrac{5 \pm \sqrt{0.36}}{2(2)}$

2. $\dfrac{5 \pm \sqrt{0.81}}{2(1)}$ $\dfrac{6 + \sqrt{0.49}}{2(4)}$ $\dfrac{-1 \pm \sqrt{0.64}}{2(5)}$

Ch 8 Ls 1 Multiply consecutive numbers. 191

_____ #1 #2 ____/ 5 #3 ____/ 4 R ___/ 6 T ____/ 15 _____
Name   Checker

#1 1. How do you change n • n = 42 to connect them? _____

2. What is the 1st step to solve n(n + 1) = 42 ? _____

3. Name 3 steps to solve it. _____

_____

#2 1. What is the trinomial?   **2 consecutive numbers multiply to get 56.**
  Use quadratics to find the 2 numbers.

How does it factor?   _____

Finish factoring it.   _____

Set both parts equal to 0.   _____

What numbers could it be?   _____   _____

_____   _____

2. What is the trinomial?   **2 consecutive odd numbers multiply to get 63.**
  Use quadratics to find the 2 numbers.

How does it factor?   _____

Finish factoring it.   _____

Set both parts equal to 0.   _____

What numbers could it be?   _____   _____

_____   _____

#3 1. $n(n + 1) = 42$ $x(x + 2) = 24$ Calculator? yes   no

Use quadritics to solve these.

2. $a(a + 4) = 32$ $b(b - 2) = 35$

Review 1. How do you change $n \cdot n = 42$ to connect them? _____   Calculator? yes   no

2. What is the 1st step to solve $n(n + 1) = 42$? _____

3. Name 3 steps to solve it. _____

4. $n(n + 5) = 50$

What is happening?

5. $a(a + 4) = 45$

6. $b(b - 3) = 40$

Ch 8 Ls 2   What starts revenue problems?   193

_____ #1 #2 ____/ 7   #3 ____/ 2   R ____/ 7   T ____/ 16 _____
              Name   Checker

#1 1. What formula does a revenue problem start with? _____
    2. Sell 20 apples for R 1 ea. What's the base equation? _____
    3. How did price and quantity change? _____
    4. How did this problem multiply same variables? _____
    5. Name 2 ways to solve it. _____
    6. What begins the formula for people getting the flu? _____
    7. What signs go in the binomials? _____
    8. How does the 2nd binomial change for people getting sick? _____
    9. How does the 1st binomial change for percent of people? _____

#2 1. Find a base   TJ sells 30 cellphones for Rs 700 each. Every Rs 20 the price
       equation.   goes up he'll sell 2 fewer. What price will make the most money?

    Change for adding Rs 20 to the price.   _____

           Where do the variables go?   _____

               What's the trinomial?   _____

           What's the highest revenue?   _____

                If you need more
                space try paper.   _____

    2. Find a base   Mr J sells 20 TVs for Rs 7000 each. Every Rs 500 the price goes
       equation.   up he'll sell 2 fewer. What price will make the most money?

    Change for adding Rs 500 to the price.   _____

           Where do the variables go?   _____

               What's the trinomial?   _____

           What's the highest revenue?   _____

#3 **Revenue Problems**

1. TJ sells 30 radios for Rs 200 each month. Every Rs 40 the price goes up he'll sell 3 fewer. What price will make the most money?   Calculator? yes no

   Find a base equation. _____

   Change for adding Rs 40 to the price. _____

   Where do the variables go? _____

   What's the trinomial? _____

   What's the highest revenue? _____

   _____

2. Ojas sold 20 hotdogs for Rs 100 an hour at his fast food stand. Each Rs 20 he raised prices he sold 2 less hotdogs. What is the best price?

   Find a base equation. _____

   Change for adding Rs 100 to the price. _____

   Where do the variables go? _____

   What's the trinomial? _____

   What's the highest revenue? _____

   _____

Review

1. What formula does a revenue problem start with? _____   Calculator? yes no
2. Sell 20 apples for R1 ea. What's the base equation? _____
3. How did price and quantity change? _____
4. How did this problem multiply same variables? _____
5. Name 2 ways to solve it. _____
6. What begins the formula for people getting the flu? _____
7. What signs go in the binomials? _____
8. How does the 2nd binomial change for people getting sick? _____
9. How does the 1st binomial change for percent of people? _____
10.   Price   Quantity   Revenue

    What's Happening? $(300 + 10d)(30 - 2d) = r$ _____
    Price Ojas sold his chicken sanwiches. _____

Ch 8 Ls 4   Quadratic Inequalities   195

_____ #1 #2 ____/ 10 #3 ____/ 4 R ____/ 7 T ____/ 21 _____
Name   Checker

#1 1. Name 3 Steps to Inequality Story Problems _____

_____ and _____

2. Name 3 steps to graph quadratic inequalities. _____

_____ and _____

3. What's happening in this equation? $y < 3x + 2$

Slope: ____   Y intercept: ____   < means _____

Make an equation for this graph.

_____

(graph with axes 0–8)

#2 1. What's the 1st step to make an inequality for a story problem? _____

2. How do you find another answer? _____

3. How do you decide which sign to use? _____

4. How can you tell if it's equal to or not? _____

5. Find a starting equation.   The water tower holds 2,000 kiloliters. It developed a leak at the bottom for 10 liters of water per hour. How long until 1/20 is almost gone? (consider that is full)

Will it use < or >?   _____

Is it equal to or not?   _____

_____

_____

#3 1.

$y < x - 2$

$y \geq 2x - 2$

Calculator? yes no

Make an equation for each inequality.

2.
$y < x + 1$

$y > \frac{1}{2}x - 1$

Review

1. Name 3 Steps to Inequality Story Problems _____

_____ and _____

2. Name 3 steps to graph quadratic inequalities. _____

_____ and _____

3. What's the 1st step to make an inequality for a story problem? _____

4. How do you find another answer? _____

5.. How can you tell if it's equal to or not? _____

6. If a car model costs Rs 20,000 and costs an average of Rs 10 per km to operate. A higher model costs Rs 25,000 with an average of Rs 8 to operate. When are they the same?

Review Problems 197

_____ #1 to #3 ____ / 10 #4 ____ / 5 T ____ / 15
Name

#1 1. Consecutive Numbers _____
    2. Revenue Problems _____
    3. Model Problems _____
    4. Inequalties _____

#2 Use quadratics to solve these.   Calculator?
                                                                    yes   no

1.   n(n + 2) = 35   x(x + 10) = 24

        _____   _____
        _____   _____
        _____   _____
        _____   _____
        _____   _____

2.   a(a + 4) = 32   b(b - 4) = 36

        _____   _____
        _____   _____
        _____   _____
        _____   _____
        _____   _____

#3 Graph Inequality Problems   Calculator?
                                                                    yes   no

1.
       y < 2x - 3   y ≥ 3x + 2

Make an equation
for each inequality.

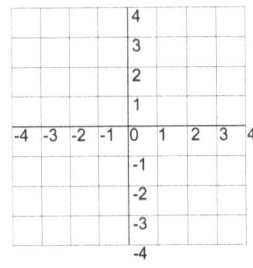
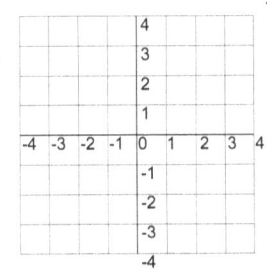

#4

Calculator? yes no

1. A band sells 1500 tickets at Rs 20 each for a raffle. For each Rs 5 increase in the ticket price, 25 fewer tickets are to be sold. What should be the ticket's price to maximize revenue?

   What will be the maximum revenue?

Write an equation before solving it.

2. The price of a music ticket is Rs 50. They sell 600 tickets. A survey found that for each Rs 5 increase in price, they think that 20 fewer tickets will be sold.

   What is the number of Rs 5 increases in price that will maximize the revenue and what is the maximum revenue?

3. A store sells an average of 400 rolls at Rs 300 per roll. Statistics show that for every Rs 2 decrease in price, customers will buy 5 additional rolls.

   What is the best price for rolls?

4. A store sells 100 winter coats in a season for Rs 500 each. Each Rs 25 decrease in the price would result in 2 more coats being sold.

   What is the best price for coats?

5. Ira is in charge of leasing apartments. They lease a 2 bedroom one for Rs 5000 a month and they have 200 to lease. They think that Rs 200 decrease in price would mean they lease 2 more apartments.

   What is the best price for renting the apartments?

Ch 9 Ls 1 How gravity formula works. 199

_____ #1 #2 ____/ 7 #3 #4 ____/ 2 R ___/ 6 T ____/ 15 _____
           Name   Checker

#1 1. What's the basic gravity formula for distance? _____
   2. How do you find how far it fell during the 2nd second? _____
   3. How much of a parabola does this graph use? _____
   4. How does the formula change to show an object thrown down? _____

#2 1. Make an equation.   **Drop a rock for 3 seconds. How far does it fall?**

   Solve it for distance.   _____

   Solve it for 4 seconds.   _____

   How far did it fall the 4th second? _____

   2. Make an equation.   **A ball is thrown down at 2 m/sec.**
                           How far will it travel in 1 second?

   Solve it for distance.   _____

   Solve it for 2 seconds.   _____

   How far did it fall the 2nd second? _____

   3. Find 3 points and graph it. _____

                                 0 1 2 3 4 5 6 7 8   seconds

#3 How gravity formula works.   Calculator? yes   no

1. A ball is thrown down at 5 m/sec. How far will it travel in 3 second?

Equation. _____

Find 1 sec. _____

Find 2 sec. _____

How far the 2nd sec? _____

2. A ball is thrown down at 10 m/sec. How far will it travel in 3 seconds?

Equation. _____

Find 1 sec. _____

Find 2 sec. _____

How far the 2nd sec? _____

3. Find 3 points and graph it.
Feet

4.
Feet

Review
1. What's the basic gravity formula for distance? _____   Calculator? yes no
2. How do you find how far it fell during the 2nd second? _____
3. How much of a parabola does this graph use? _____
4. How does the formula change to show an object thrown down? _____

What is happening.

5. $h = -5(4)^2$ sec

6. $h = -5(3)^2 - 10(3)$ sec
Find how far a rock fell.

Ch 9 Ls 2 How gravity formula solves for height.   201

_____ #1 #2 ____/ 6 #3 ____/ 5 R ___/ 8 T ____/ 19 _____
             Name   Checker

#1   1. How can gravity formula solve for height instead of time? _____
     2. What does H stand for? _____
     3. What is gravity formula with starting height and a push? _____

    #2   1. Make an equation.   **A jar falls 60 meters from the top of a cliff.**
                                 How long until it hits the ground below?

              Solve it for distance.   _____

                                              _____

                                              _____

         2. Make an equation.   **A coin drops from a 100 meter building.**
                                 How long until it hits the ground?

              Solve it for distance.   _____

                                              _____

                                              _____

         3. Find 2 points and graph it.   _____

                                    0 1 2 3 4 5 6 7 8   seconds

#3

1. A rock falls 48 meters from the top of a cliff. How long until it hits the ground? Equation. _____ Calculator? yes no

Check to see how long it takes these to fall.

Solve it for how long it takes to fall. _____

How far did it fall the 4th second? _____

2. A wallet drops from a 96 meter building. How long until it hits the ground? Equation. _____

Solve it for how long it takes to fall. _____

How far did it fall the 2nd second? _____

3. A rock falls 140 meter from the top of a cliff. How long until it hits the ground? Equation. _____

Solve it for how long it takes to fall. _____

How far did it fall the 4th second? _____

4. A stone drops from a 200 meter building. How long until it hits the ground? Equation. _____

Solve it for how long it takes to fall. _____

How far did it fall the 2nd second? _____

5. A coin drops from a 600 meter building. How long until it hits the ground? Equation. _____

Solve it for how long it takes to fall. _____

How far did it fall the 2nd second? _____

Review 1. How can gravity formula solve for height instead of time? _____

2. What does H stand for? _____

3. What is gravity formula with starting height and a push? _____

Ch 9 Ls 3 Gravity object thrown up and down.   203

_____ #1 #2 ____ / 6 #3 ____ / 4 R ___ / 6 T ____ / 16 _____
        Name   Checker

#1 1. How does gravity formula change for an object thrown up? _____
        2. Why is velocity added instead of subtracted? _____
        3. What part of the equation starts it? _____
        4. When would the ending height be 0? _____

#2 1. What is starting and ending height? $0 = -5t^2 + 60t + 50$

                How high is it after 1 second?   Start: ____ m   End: ____ m

                How high is it after 2 seconds? _____

                What is the highest point?   _____

                How long until it touches down? _____

                                                 _____

                                                 _____

        2. What is starting and ending height? $0 = -5t^2 + 12t + 15$

                How high is it after 1 second?   Start: ____ m   End: ____ m

                How high is it after 2 seconds? _____

                What is this the highest point? _____

                How long until it touches down? _____

                                                 _____

                                                 _____

#3 Solve these problems using the quadratic distance problem.   Calculator?
                                                                   yes   no

1. Throw a baseball up at 40 m/sec.
   How far does it go in 3 seconds?   _____

                                      _____

                                      _____

2. Throw a futball up at 10 m/sec.
   How far does it go at it's highest _____
   level?
                                      _____

                                      _____

3. Hit a baseball up at 40 m/sec. How
   far does it go in 4 sec?   _____

                                      _____

                                      _____

4. Kick a futball up at 20 m/sec. How
   far does it go in 5 seconds?   _____

                                      _____

                                      _____

Review 1. How does gravity formula change for an object thrown up? _____   Calculator?
                                                                                      yes   no
2. Why is velocity added instead of subtracted? _____

3. What part of the equation starts it? _____

4. When would the ending height be 0? _____

5. $0 = -5t^2 + 80t + 10$ _____

Solve it.   _____

6. $10 = -5t^2 + 20t + 30$ _____

                             _____

Review Problems   205

_____   #1 to #3 ____ / 9   #4 #5 ____/11 Total ____/ 20
           Name

1. Quick Formula _____
2. Real Graph _____
3. Graviiy Formula _____
4. Thrown Objects _____
5. Thrown Up/Down _____
6. Different Heights _____

#2 Design the equation to fit the graph.   Calculator?
                                                  yes   no

1. **Vertex 1, 2   Y int: 6   Opens Down**

   Start equation.   _____
   Find the A term.   _____
   Solve the 2nd step. _____
   Find the fraction. _____
   New equation   _____

2. **Vertex -2, -3   Y int: 2   Opens Up**

   Start equation.   _____
   Find the A term.   _____
   Solve the 2nd step. _____
   Find the fraction. _____
   New equation   _____

#3 Find the B Terms   Calculator?
                                                  yes   no

1. $y = 4x^2 + x + 20$ Find the y term. _____

   $\dfrac{-b}{2a}$ $\dfrac{}{2}$ = x   What's next? _____

            = x   _____

            = x   _____ vertex ____, ____

#4 Solve using the quadratic formula.   Calculator? yes no

1. $d = -\frac{1}{2} 32(4)^2$ ft/sec² $d = -\frac{1}{2} 10(8)^2$ m/sec² $d = -5(6)^2$ m/sec²

   _____   _____   _____

   _____   _____   _____

2. $d = -16(8)^2$ ft/sec² $d = -5(8)^2$ m/sec² $d = -16(12)^2$ ft/sec²

   _____   _____   _____

   _____   _____   _____

3. $d = -\frac{1}{2} 32(6)^2$ ft/sec² $d = -\frac{1}{2} 10(14)^2$ m/sec² $d = -5(7)^2$ m/sec²

   _____   _____   _____

   _____   _____   _____

#5 Solve using the quadratic formula.   Calculator? yes no

1. **Vertex 1, 3   Y int: 4   Opens Down**

   Start equation. _____

   Find the A term. _____

   Solve the 2nd step. _____

   Find the fraction. _____

   New equation _____

2. **Vertex -2, -1   Y int: 10   Opens Up**

   Start equation. _____

   Find the A term. _____

   Solve the 1st step. _____

   Find the fraction. _____

   New equation _____

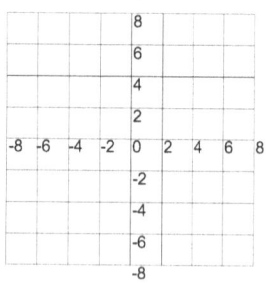

Review Problems 207

_____ Name Front ____/5 Back ____/5 Total ____/10

Calculator? yes no

#1. 1. A relief package is released from a helicopter at 1200 meters. The height of the package can be modeled by the equation h = -5t^2 + 1200.

Make an equation, solve it.

The pilot wants to know how long it will take for the package to hit the ground.

2. The height of a rocket launched upward from a 40 meter cliff is modeled by h = 5t^2 + 32t + 144. How long will it take to land?

3. Bob threw a rock off a bridge into the river. It's 20 meters high. We use h = -5t^2 - 15t + 20 for an equation. How long did it take the rock to hit the surface of the water?

4. During a game of golf, Kate hits her ball out of a sand trap. We use the equation h = -5t^2 + 10t - 2. How long did it take Kate's ball to hit the ground?

5. An object is dropped and it takes 3 seconds to hit the ground. What's the original height of the object?

#2

Calculator? yes no

1. A rock falls 80 meters from the top of a cliff. How long until it hits the ground?

2. A ball bounces straight upward to a maximum height of 3 meters. How long will it take to reach maximum height?

3. The profit a manufacturer makes each day is labeled $P < -x^2 + 140x - 1200$, where P is the profit and x is the price sold. For what values of x does the company make a profit?

4. A baseball is hit by a batter. The height of the ball, H, is shown by an equation $h = -5t^2 + 20t + 1$. What is the highest point and how far does it go?

5. A person stands on top of a 100 meter building throws a stone vertically upward. The equation is $h(t) = -5t^2 + 30t + 100$ where h(t), in feet, t sec after it was thrown. After how many seconds does the ball reach its maximum height? Round to the nearest unit.

Ch 10 Ls 1 Use the Vertex Equation.   209

_____ #1 #2 ____ / 7 #3 ____ / 4 R ___ / 9 T ____ / 20 _____
         Name   Checker

#1 1. There's 3 parts to the Vertex Formula. What does **a** show? _____
    2. What do **h** and **k** show? _____
    3. If the vertex is at **4, 6** what would the equation be? _____

#2 1. Decide what's happening. $y = (x - 2)^2 - 1$
        Plot a quick graph.

    Steep **less than, greater than, equal to**

    Opens **Up Down**   vertex is _____

    2. Decide what's happening. $y = 2(x - 3)^2 + 3$
        Plot a quick graph.

    Steep **less than, greater than, equal to**

    Opens **Up Down**   vertex is _____

    3. Make a quick equation. How steep is   **Slope: 4   Vertex is at: (1, 2)**
        it? How does it move? Up or down?   **Graph: Down**

        _____

    4. How steep is it? How does it move?   **Slope: 2   Vertex is at: (-2, 2)**
                                             Graph: Up

        _____

#3 Find the matching graphs for each equation.   Calculator? yes   no

1. $y = -(x^2 - 2) + 0$

Opens Up   Down
vertex is _____

X intercepts
yes   no

$y = -2(x^2 - 1) - 1$

Opens Up   Down
vertex is _____

X intercepts
yes   no

2. $y = 2(x + 2)^2 - 4$

Opens Up   Down
vertex is _____

X intercepts
yes   no

$y = -3(x - 4)^2 - 1$

Opens Up   Down
vertex is _____

X intercepts
yes   no

Review 1. There's 3 parts to the Quick Formula. What does **a** show? _____   Calculator?
2. What do h and k show? _____   yes   no
3. If the vertex is at **4, 6** what would the equation be? _____

Make equations for these quick formulas.

4. Slope: -2 Vertex: -1, 2   Slope: 4 Vertex: 2, 3

    _____   _____

5. Slope: 1 Vertex: 1, -3   Slope: $\frac{1}{2}$ Vertex: -1, 4

    _____   _____

6. Slope: -3 Vertex: 4, 5   Slope: 2 Vertex: -3, 2

    _____   _____

Ch 10 Ls 2 Make a real graph. 211

_____ #1 #2 ____ / 6 #3 ____ / 3 R ____ / 8 T ____ / 18 _____
                Name   Checker

#1 1. What is a Real Graph? _____

2. What part moves the Quick Formula up or down? _____

3. What part moves the Quick Formula side to side? _____

4. What part decides the size of the parabola? _____

#2 1. Find the start equation.   **A parabola is 50 meters tall and 30 meters wide.**

_____   How do you find the A term?

_____   Solve the 1st step.

_____   What fraction is the A term?

_____   Use another line if needed.

_____   What is the equation?

_____

2. Find the start equation.   **A parabola is 20 meters tall and 40 meters wide.**

_____   How do you find the A term?

_____   Solve the 1st step.

_____   What fraction is the A term?

_____   Use another line if needed.

_____   What is the equation?

_____

#3 1. A golfer drives 120 meters with a vertex of 40 meters. What is the equation?

Start equation. _____   Calculator?
Find the A term. _____   yes   no
Solve the 2nd step. _____
Find the fraction. _____
New equation _____

2. A soccer player shoots a straight shot from 20 meters. and it's vertex is 5 m high. What's the equation?

Start equation. _____
Find the A term. _____
Solve the 2nd step. _____
Find the fraction. _____
New equation _____

3.

$y = 3x^2$

Base Rate
_____

$y = 5x^2$

Base Rate
_____

Find each equation's Base Rate. Graph.

4. **Vertex 0, -4   Y int: 2   Opens Up**

Start equation. _____

Find the A term. _____
Solve 2nd step. _____
Find the fraction. _____
New equation _____

Review   1. What is a Real Graph? _____
2. What part moves the Quick Formula up or down? _____
3. What part moves the Quick Formula side to side? _____
4. What part decides the size of the parabola? _____

Ch 10 Ls 3 Use Quick Equation with graphs. 213

_____ #1 #2 ____ / 6 #3 ____ / 3 R ____ / 4 T ____ / 13 _____
         Name   Checker

#1 1. What equation shows a parabola that's 30 feet tall and opens down? _____

_____

2. What makes the parabola 50 ft wide? _____

3. How do you find the A term? _____

4. What does the A term show? _____

#2 1. Find the start equation.   **Take off at 8 m, vertex at (30, 40), and lands at 20 m.**

_____   How do you find the A term?

_____   Solve the 1st step.

_____   What fraction is the A term?

_____   Use another line if needed.

_____   What is the equation?

_____

2. Find the start equation.   **Take off at 10 m, vertex at (40, 44), and lands at 30 m.**

_____   How do you find the A term?

_____   Solve the 1st step.

_____   What fraction is the A term?

_____   Use another line if needed.

_____   What is the equation?

_____

#3 1. **Y int: 2   Opens Down   Vertex 2, 4**   Calculator? yes   no

   Start equation. _____

   Find the A term. _____

   Solve the 1st step. _____

   Find the fraction. _____

   New equation _____

2. **Y int: 6   Opens Up   Vertex 0, 6**

   Start equation. _____

   Find the A term. _____

   Solve the 1st step. _____

   Find the fraction. _____

   New equation _____

3. **Y int: 4   Opens Down   Vertex 1, 5**

   Start equation. _____

   Find the A term. _____

   Solve the 1st step. _____

   Find the fraction. _____

   New equation _____

Review   1. What equation shows a parabola that's 30 feet tall and opens down? _____

   _____

   2. What makes the parabola 50 ft wide? _____

   3. How do you find the A term? _____

   4. What does the A term show? _____

Review Problems 215

_____ #1 #2 ____ / 10 #3 ____ / 8 Total ____ / 18
         Name

1. Quick Equations _____

2. Real Graphs _____

#2 Make a graph for each equation.

Calculator?

1. $y = -(x^2 + 4) + 1$ $y = -(x^2 - 2) - 3$ yes   no

Opens Up   Down   Opens Up   Down
vertex is _____   vertex is _____

X intercepts   X intercepts
  yes   no   yes   no

2. $y = 3(x + 2)^2 - 4$ $y = -4(x - 3)^2 - 2$

Opens Up   Down   Opens Up   Down
vertex is _____   vertex is _____

X intercepts   X intercepts
  yes   no   yes   no

3. $y = 2(x + 4)^2 - 2$ $y = -2(x - 4)^2 - 1$

Opens Up   Down   Opens Up   Down
vertex is _____   vertex is _____

X intercepts   X intercepts
  yes   no   yes   no

#3

Make an equation and graph it.

1. A kick ball player kicks it 40 m/s and it's vertex is 10 meters. Equation? Graph?

   Calculator? yes no

   _____

   _____

   _____

   _____

   _____

2. A soccer player kicks the ball 60 maters with vertex of 20 meters. Equation? Graph?

   _____

   _____

   _____

   _____

3. A shot putter throws 32 m from 2 meters. The vertex is 8 meters. Equation? Graph?

   _____

   _____

   _____

   _____

4. A punter punts the ball 60 meters with a vertex of 50 meters. What's the equation? Graph?

   _____

   _____

   _____

   _____

   Time to take the quiz.

Ch 11 Ls 1 Quadratic Distance Formula 217

_____ #1 #2 ____ / 7 #3 ____ / 9 R ____ / 7 T ____ / 23 _____
Name   Checker

#1 1. What is the quadratic formula? _____
    2. How does the formula change for a trip already in progress? _____
    3. Does it use total or unit acceleration? _____

#2 1. Make an equation.   Accelerate at 4 m/sec^2 from 0 for 5 seconds. How far did it go?

    Solve for distance.

  2. Make an equation.   Accelerate at 9 m/sec^2 from 0 for 10 seconds. How far did it go?

    Solve for distance.

  3. What's happening in this equation? $d = \frac{1}{2} 6(8)^2$ m/sec^2

    Solve for distance.
    First step?   Accelerate ____ m/s/s in ____ sec. Find _____.

    Finish it.

  4. What's happening in this equation? $d = \frac{1}{2} 8(10)^2$ m/sec^2

    Solve for distance.
    First step?   Accelerate ____ m/s/s in ____ sec. Find _____.

    Finish it.

#3 Solve these equations with quadratic distance formula.   Calculator? yes no

1. $d = \frac{1}{2} 2(5)^2$ m/sec² $d = \frac{1}{2} 8(5)^2$ m/sec² $d = \frac{1}{2} 6(8)^2$ m/sec²

Go ___ m/s/s in ___ sec.   Go ___ m/s/s in ___ sec   Go ___ m/s/s in ___ sec

Accelerate_____   Accelerate_____   Accelerate_____

2. $d = \frac{1}{2} 12(7)^2$ m/sec² $d = \frac{1}{2} 10(5)^2$ m/sec² $d = \frac{1}{2} 14(20)^2$ m/sec²

Go ___ m/s/s in ___ sec.   Go ___ m/s/s in ___ sec   Go ___ m/s/s in ___ sec

Accelerate_____   Accelerate_____   Accelerate_____

3. $d = \frac{1}{2} 16(15)^2$ m/sec² $d = \frac{1}{2} 18(9)^2$ m/sec² $d = \frac{1}{2} 20(8)^2$ m/sec²

Go ___ m/s/s in ___ sec.   Go ___ m/s/s in ___ sec   Go ___ m/s/s in ___ sec

Accelerate_____   Accelerate_____   Accelerate_____

Review 1. How does rate find distance? _____   Calculator?
2. How does acceleration find distance? _____   yes no
3. What changes if the acceleration starts with a trip? _____
4. Does it use total or unit acceleration? _____

Find these.

5. $60 = \frac{1}{2} 10t^2$ m/sec² $120 = \frac{1}{2} 20t^2$ m/sec² $200 = \frac{1}{2} 40t^2$ m/sec²

___ m in ___ m/s/s.   ___ m in ___ m/s/s   ___ m in ___ m/s/s

Time_____   Time_____   Time_____

Ch 11 Ls 2 Quadratic distance slows down. 219

_____ #1 #2 ____ / 6 #3 ____ / 2 R ___/ 5 T ____/ 13 _____
        Name   Checker

#1 1. How does Half AT squared change for slowing down? _____
    2. What is happening? $40 = \frac{1}{2} - 5t^2 + 30t$ _____
    3. Name 3 ways to solve an equation. _____
    _____

#2 1. Make an equation. **A truck is going 20 m/sec. A rabbit is 80 m ahead. If the truck slows down at 10 m/sec^2, how long until it stops?**

    Solve for distance. _____

    2. Make an equation. **A car is going 30 m/sec. A rabbit is 50 m ahead. If the car slows down at 5 m/sec^2, how long until it stops?**

    Solve for distance. _____

    3. What's happening in this equation? $240 = \frac{1}{2} - 10t^2 + 40t$
                                                          m/sec²

    Deaccelerate _____ m/s/s in _____ m traveling _____ m/sec. Find _____.

#3 Solve these problems with quadratic distance formulas.   Calculator? yes no

1. A subway is able to go from 0 to 30 m/sec traveling at 5 m/sec/sec, how long will it take?

   Equation _____
   1st step? _____
   Next step? _____
   Last step? _____
   Answer. _____
   Other answer? _____

2. A roller coaster is able to also go 30 m/sec, but it takes 5 seconds to stop. How many m/sec/sec will it take to stop?

   Equation _____
   1st step? _____
   Next step? _____
   Last step? _____
   Answer. _____
   Other answer? _____

Review 1. How does Half AT squared change for slowing down? _____   Calculator? yes no

2. What is happening? $40 = \frac{1}{2}(-5)t^2 + 30t$ _____

3. Name 3 ways to solve an equation. _____
   _____

What's Happening?

4. $160 = \frac{1}{2} \cdot 10t^2 + 20t$
   m/sec²

   _____
   _____
   _____

5. $0 = \frac{1}{2} 10(7)^2 + 40(7)$
   m/sec²

   _____
   _____

Ch 11 Ls 3 Connect Gravity and Acceleration   221

_____ #1 #2 ____/ 10 #3 ____/ 12 R ___/ 4 T ____/ 26 _____
          Name   Checker

#1 1. What acceleration formula doesn't use time? _____

2. What is the gravity formula for meters? _____

3. How does the moon's gravity compared to earth's? _____

4. Which planets are about earth's gravity? _____

5. Make an equation — Venus, Saturn, and Uranus gravity is like Earth's. Make an equation for gravity.

$$d = \underline{\hspace{3cm}}$$

6. Make an equation — Mars and Mercury gravity is 1 3rd of Earth's. Make a gravity equation for it.

$$d = \underline{\hspace{3cm}}$$

7. Make an equation — Jupiter's Gravity is 2.6 times of Earth's. Make an equation for gravity.

$$d = \underline{\hspace{3cm}}$$

#2 1. Metric or English? What's happening? $d = \frac{1}{2} - (32)t^2$

    Metric   English   _____

2. Metric or English? What's happening? $d = -(16)t^2$

    Metric   English   _____

3. Metric or English? What's happening? $d = \frac{1}{6} - (32)t^2$

    Metric   English   _____

2. Metric or English? What's happening? $d = -5t^2$

    Metric   English   _____

#3 What is the distance for each one?   Calculator? yes no

1. $d = -\frac{1}{2}10(2)^2$ m/sec² $d = -\frac{1}{2}10(5)^2$ m/sec² $d = -5(3)^2$ m/sec²

_____ _____ _____

_____ _____ _____

2. $d = -5(5)^2$ m/sec² $d = -5(7)^2$ m/sec² $d = -4(8)^2$ m/sec²

_____ _____ _____

_____ _____ _____

3. $-64 = -5t^2$ m/sec² $60 = -5t^2$ m/sec² $-96 = -5t^2$ m/sec²

_____ _____ _____

_____ _____ _____

4. $-80 = -5t^2$ m/sec² $-40 = -5t^2$ m/sec² $-144 = -5t^2$ m/sec²

_____ _____ _____

_____ _____ _____

Review   1. How does gravity formula follow acceleration formula? _____

           2. What is the gravity formula for meters? _____

           3. How does the moon's gravity compared to earth's? _____

           4. Which planets are about earth's gravity? _____

Ch 11 Ls 4 Acclelration/Jet slow down with quadratic 223

_____ #1 #2 ____/9 #3 ____/2 R ___/10 T ____/21 _____
         Name   Checker

#1 1. What's the Acceleration Formula to find Rate? _____
    2. What happens if it adds a rate? _____
    3. What's the 1st formula without the square root? _____
    4. Make an equation. **Accelerate at 2 m/sec/sec for 100 m.**

                          _____ Solve for final speed.

    5. Make an equation. **Accelerate at 8 m/sec/sec for 50 m.**

                          _____ Find final speed.

#2 1. What formula finds how long it took to stop? _____
    2. You know how long it took. What formula finds distance? _____
    3. What 2 formulas do they use to find how long the jet takes to stop? _____

    4. Find 1st equation.   **A passenger jet is going 120 m/sec as it slows down at 10 m/sec/sec. Is a 700 m runway long enough**

           How many seconds to stop?   y = ___ x _____
           What's the 1st equation?   _____
           Does it slow down in time?   Sketch
                                                           a graph.

#3
1. A military jet is going 100 m/sec as it slows down at 15 m/sec/sec. Is a 500 m runway long enough?

2. A 4 seater local transport is going 40 m/sec as it slows down at 10 m/sec/sec. Is a 400 m runway long enough?   Calculator? yes   no

Solve these.

Equation _____

1st step? _____

Next step? _____

Last step? _____

Answer. _____

Other answer? _____
_____

Equation _____

1st step? _____

Next step? _____

Last step? _____

Answer. _____

Other answer? _____
_____

Review
1. What's the Acceleration Formula to find Rate Formuls? _____   Calculator? yes   no
2. How does the formula change if it starts from a rate? _____
3. What's the 1st formula without the square root? _____
4. What formula finds how long it took to stop? _____
5. You know how long it took. What formula finds distance? _____
6. What 2 formulas do they use to find how long the jet takes to stop? _____

Solve these.

7. $\sqrt{2(10)8} = r_f$
    m/sec² m

$\sqrt{2(30)10} = r_f$
m/sec² m

_____

_____

_____

_____

8. $\sqrt{2(200)5} = r_f$
    m/sec²   m

$\sqrt{2(400)7} = r_f$
mt/sec²   m

_____

_____

_____

_____

Review Problems 225

_____   #1 #2 #3 _____ / 18 #4 _____ / 5   T _____ / 23
          Name

#1   1. Quadratic Distance formula _____

     2. Time Braking _____

     3. Rate from Acceleration _____

#2   Solve these equations with quadratic distance formula.   Calculator?
                                                                      yes   no

1. $d = \frac{1}{2} 3(7)^2$ $d = \frac{1}{2} 7(8)^2$ $d = \frac{1}{2} 5(9)^2$
           m/sec²   m/sec²   m/sec²

     Go ____ m/s/s in ____ sec.   Go ____ m/s/s in ____ sec   Go ____ m/s/s in ____ sec

     Acc _____   Acc _____   Acc _____

2. $d = \frac{1}{2} 10(11)^2$ $d = \frac{1}{2} 9(12)^2$ $d = \frac{1}{2} 15(20)^2$
           m/sec²   m/sec²   m/sec²

     Go ____ m/s/s in ____ sec.   Go ____ m/s/s in ____ sec   Go ____ m/s/s in ____ sec

     Acc _____   Acc _____   Acc _____

3. $d = \frac{1}{2} 14(15)^2$ $d = \frac{1}{2} 17(8)^2$ $d = \frac{1}{2} 20(5)^2$
           m/sec²   m/sec²   m/sec²

     Go ____ m/s/s in ____ sec.   Go ____ m/s/s in ____ sec   Go ____ m/s/s in ____ sec

     Acc _____   Acc _____   Acc _____

#3 1. $d = -\frac{1}{2} 10(3)^2$ $d = -\frac{1}{2} 20(6)^2$ $d = -5(4)^2$ Calculator?
              m/sec²   m/sec²   m/sec²   yes   no

Solve using the   _____   _____   _____
quadratic formula. _____   _____   _____

    2. $d = -5(10)^2$ $d = -5(8)^2$ $d = -5(7)^2$
              m/sec²   m/sec²   m/sec²

         _____   _____   _____
         _____   _____   _____

#4 Story Problems.

1. A russian jet is going 120 m/sec as it slows down at 20 m/sec/sec. Is a 500 m runway long enough?

Calculator? yes   no

2. A local plane transport is going 30 m/sec. It slows down at 8 m/sec/sec. Is a 1000 m runway long enough for the plane?

3. A subway is able to reach to reach 40 m/sec. The subway is able to stop traveling -10 m/s/s. How long will it take to stop?

4. A truck is going 26 m/sec when a tree falls 35 meters in front of it. The truck slows down at 10 m/sec/sec. Is the truck able to stop in time?

5. Ojas is driving a car going 30 m/sec. He wants to pass a slower car also going 30 m/sec. If he accelerates at 5 m/sec/sec, how long will it take the slower car?

Ch 12 Ls 1 Solve fractions in equations.   227

_____ #1 #2 ____/ 6   #3 ____/ 4   R ____/ 4   T ____/ 14   _____
Name   Checker

Example $\dfrac{1}{5} + \dfrac{1}{2} = \dfrac{x}{4}$

#1   1. What's 1st to change all of an equation's denominators? _____
     2. What's the 2nd step? _____
     3. What's the last step? _____
     4. How can you remember the 3 steps? _____

#2   1. Write the 3 denominators above them. What's left?

$\overset{6\times 3\times 2}{\dfrac{1}{6}} + \overset{6\times 3\times 2}{\dfrac{1}{3}} = \overset{6\times 3\times 2}{\dfrac{x}{2}}$

_____ Solve the 1st step of FOX.

_____ Solve the 2nd step of FOX.

_____ X intercepts?

_____

2. Write the 3 denominators above them. What's left?

$\overset{a\ a+4\ 3}{\dfrac{1}{a}} + \overset{a\ a+4\ 3}{\dfrac{1}{a+4}} = \overset{a\ a+4\ 3}{\dfrac{2}{3}}$

_____ Solve the 1st step of FOX.

_____ Solve the 2nd step of FOX.

_____ X intercepts?

_____

#3

Solve these with WCM.
Round it.

1. 7 4 2   7 4 2   7 4 2
   $\frac{5}{7} + \frac{3}{4} = \frac{x}{2}$

   6 5 4   6 5 4   6 5 4
   $\frac{5}{6} + \frac{2}{5} = \frac{x}{4}$

2. n n+2 7   n n+2 7   n n+2 7
   $\frac{1}{n} + \frac{1}{n+2} = \frac{1}{6}$

   b 3b 4   b 3b 4   b 3b 4
   $\frac{1}{b} + \frac{1}{3b} = \frac{1}{4}$

Example $\frac{1}{5} + \frac{1}{2} = \frac{x}{4}$

Review 1. What's 1st to change all of an equation's denominators? _____

2. What's the 2nd step? _____

3. What's the last step? _____

4. How can you remember the 3 steps? _____

Ch 12 Ls 2 Work Problems. 229

_____   #1 #2 ____ / 6   #3 ____ / 1   R ____ / 11   T ____ / 18 _____
Name   Checker

#1 1. How does a work problem use a binomial? _____
2. What is the key to work problems? _____
3. How do you get same denominators for an equation? _____
4. How did it factor out? _____

Find the rate formulas, then solve the work problem.

#2 1. What are the equations?

Together, Bill and Dave mow a lawn in 6 hours. Alone, Bill needs 5 hrs longer than Dave. Find each one.

Bill ____ X ____ = ____   Dave ____ X ____ = ____   Solve each rate.

Make an equation.
What does it equal?

Write the
equation left over.

Solve the 1st
step of FOX.

Find the X intercepts.

How long did it
take each one?

2. What is happening?

$$\text{Hours} \quad \underset{\text{Dave's}}{\frac{1}{t}} + \underset{\text{Bill's}}{\frac{1}{t+5}} = \underset{\text{Together}}{\frac{1}{6}}$$

_____

#3 1. Amav can paint a room in 4 hours. Mitul would take 7 hours. How long would it take for them together?

Solve these with WCM.

$$7\,4\,x \quad 7\,4\,x \quad 7\,4\,x$$
$$\frac{1}{7} + \frac{1}{4} = \frac{1}{x}$$

Calculator? yes no

2. One pipe can fill a pool 1.5 times as fast as a 2nd pipe. When both pipes are opened, they fill the pool in five hours. How long would it take to fill the pool if only the slower pipe is used?

$$x\ 2x\ 5 \quad x\ 2x\ 5 \quad x\ 2x\ 5$$
$$\frac{1}{x} + \frac{3}{2x} = \frac{1}{5}$$

3. Working together, Mr T and Adah painted a fence in 8 hours. Last year, Mr T painted the fence by himself. The year before, Adajh painted it by herself, but took 12 hours more than Mr T took. How long did Mr T and Adah take, when each was painting alone?

$$n\ n-12\ 8 \quad n\ n-12\ 8 \quad n\ n-12\ 8$$
$$\frac{1}{n} + \frac{1}{n-12} = \frac{1}{8}$$

Review 1. How does work problem use a binomial? _____

2. What is the key to work problems? _____

3. How do you get same denominators for an equation? _____

4. How did it factor out? _____

5. What is the 1st step to the river problem? _____

6. How does 8 kph current change each formula? _____

7. What does each rate formula equal? _____

Calculator? yes no

Ch 12 Ls 3 Stream Problems   231

_____   #1 #2 ____/ 6 #3 ____/ 1 R ____/ 6 T ____/ 13 _____
Name   Checker

#1 1. What do stream problems solve for? _____
    2. How does rate formula change for 8 mph slower? _____
    3. What are the rate formulas? _____
    4. What are 4 steps to solve stream problems? _____

#2 1. What are the   A boat travels 10 kilometers against an 4 mph current and 10 km
    equations?   with a current. It's a 2 hours round trip. Find speed in calm water.

    With ____ X ____ = ____   Against ____ X ____ = ____   Solve each rate.

                                                   Make an equation.
                                                   What does it equal?

                                                   Write the
                                                   equation left over.

                                                   Solve the 1st
                                                   step of FOX.

                                                   Find the X intercepts.

                                                   How long did it
                                                   take each one?

                                        Trip #1   Trip #2   Hours
    2. What is happening?   Miles $\dfrac{30}{r+8} + \dfrac{30}{r-8} = \dfrac{2}{1}$
                            Mph

#3 1. A boat goes downstream and covers the distance between two ports in 4 hrs., while it covers the same distance upstream in 5 hrs. If the speed of the stream is 2km/h, find the speed of the steamer in still water.

____ x ____ = ____

____ x ____ = ____

Calculator? yes   no

2. Ojas traveled against the wind in a plane for 3.3 hr. The return trip with the wind took 2.7 hr. Find the speed of the wind if the speed of the plane is still air is 220 kph.

____ x ____ = ____

____ x ____ = ____

3. Superman flies 125 kilometers with a head wind in 216 seconds and back in 210 sec. What is his speed without wind?

____ x ____ = ____

____ x ____ = ____

Review   1. What do stream problems solve for? _____

2. How does the rate formula change for the time? _____

3. What is each rate formula? _____

4. What are 4 steps to solve stream problems? _____

Calculator? yes   no

Ch 12 Ls 4 Quadratic Fence Problems   233

_____   #1 #2 ___ / 7 #3 ___ / 8 R ___ / 4 T ___ / 19 _____
        Name   Checker

#1 1. What is this lesson about? _____

2. What is the equation for this rectangle? w (w + 2) _____

3. What is the difference between area and perimeter? _____

_____

4. What solved this perimeter problem? _____

5. What solved the area problem? _____

#2 1. What is the equation?   CJ wants to build a garden so that three sides are fenced and the 4th side is a fence. He has 45 meters of fencing available. Find the maximum enclosed area. (The first equation is important.)

_____ X _____ = ____   Solve the equation.

2. What is the equation?   You have a 360 meter roll of fencing and a large field. You want to construct a rectangular playground area. What are the dimensions of the largest such yard? What is the largest area?

_____ X _____ = ____   Solve the equation.

#3 The First Equation is Important

1. Pedro has 240 m of fencing to put around a field that has an area of 3600 sq m. Which equation can be used to find the length of the field?

2. Mr K wants to enclose a rectangular corral next to the barn. The side of the barn will form one side of it. The other three sides will be fencing. Mr K purchased 120 m of fencing and wishes to enclose an area of at least 1600 sq m.

3. CJ has 90 m of fencing to build a fence for his dog. He plans to build the pen using one side of a 30 m long building. He will use all of the fencing for the other three sides of the pen. Use the area function for this rectangle to determine the area of the pen.

4. Mitul purchased 400 m of fencing to build a corral for his horses. Each horse requires 1000 m^2 of space, what is the maximum number of horses Mitul can put in the corral?

Review 1. What is the equation for this rectangle? $w(w + 2)$ _____
        2. What is the difference between area and perimeter? _____

        3. What solved this perimeter problem? _____
        4. What solved the area problem? _____

Review Problems 235

_____ #1 #2 ____ / 7 #3 ____ / 5 total ____ / 12
Name

#1 1. **Same Denominators for an Equation** _____

_____

2. **Work Problems** _____
3. **Stream Problems** _____
4. **Fence Problems** _____

#2 1.

5 4 6   5 4 6   5 4 6
$$\frac{3}{5} + \frac{1}{4} = \frac{x}{6}$$

8 5 3   8 5 3   8 5 3
$$\frac{3}{8} + \frac{2}{5} = \frac{x}{3}$$

_____ _____

_____ _____

_____ _____

_____ _____

_____ _____

2.

n n+2 7   n n+2 7   n n+2 7
$$\frac{1}{n} + \frac{1}{n+2} = \frac{2}{7}$$

a a+3 4   a a+3 4   a a+3 4
$$\frac{1}{a} + \frac{1}{a+3} = \frac{3}{4}$$

_____ _____

_____ _____

_____ _____

_____ _____

#3

Calculator? yes no

1. An aircraft carrier made a trip to Guam and back. The trip there took seven hours and the trip back took nine hours. It averaged 8 km/h on the return trip. Find the average speed of the trip there.

2. Together, Ojas and Amav mow a lawn in 6 hours. Alone, Ojas needs 5 hrs longer than Amav. Find each one.

3. A small airplane with a head wind travels 1,200 kilometers from airport A to B in 11 hours and back in 7 hours. What is the speed with no wind?

4. One pipe can fill a pool 1.5 times faster than a second pipe. When both pipes are opened, they fill the pool in five hours. How long would it take to fill the pool if only the faster pipe is used?

5. A passenger plane made a trip to Las Vegas and back. On the trip there it flew 440 kph and on the return trip it went 480 kph. How long did the entire trip there take if the return trip took 4 hours?

Ch 13 Ls 1 All the Money Interest Equations. 237

_____ #1 #2 ____ / 8 #3 ____ / 10 R ___ / 4 T ____ / 22 _____
Name   Checker

#1 1. What interest formula finds all the money? _____

   2.   Make an equation.   **Rs 20,000 at 10% interest for 1 year.**

        Solve it for all the money.

   3.   Make an equation.   **Rs 40,000 at 8% interest for 1 year.**

        Solve it for all the money.

#2 1. How does interest formula change for more than 1 year? _____
   2. How do you say a compound interest formula? _____
   3. Why is compound interest different from simple? _____

   4. Make an equation.   **Rs 50,000 at 12% interest for 2 years.**

        Solve it for all the money.

   5. Make an equation.   **Rs 60,000 at 10% interest for 3 years.**

        Solve it for all the money.

#3 Decide what's happening, then solve these rupee questions.   Calculator? yes no

1. 4000 x 1.08 = t   5000 x 1.06 = t
    ____ at ___ interest for 1 yr   ____ at ___ interest for 1 yr

    _____   _____

2. 9000(1 + 1.50)² = t   12000(1 + 1.60)² = t
    ____ at ___ interest for ___ years   ____ at ___ interest for ___ years

    _____   _____

3. 7000(1 + 1.20)² = t   3000(1 + 1.10)² = t
    ____ at ___ interest for ___ years   ____ at ___ interest for ___ years

    _____   _____

4. 8000(1 + 1.40)⁴ = t   2000(1 + 1.70)³ = t

You can use a calculator.
    ____ at ___ interest for ___ years   ____ at ___ interest for ___ years

    _____   _____

5. 9000(1 + 1.60)⁵ = t   3000(1 + 1.50)⁶ = t
    ____ at ___ interest for ___ years   ____ at ___ interest for ___ years

    _____   _____

Review 1. What interest formula finds all the money? _____
        2. How can you write the interest formula for less than 100%? _____
        3. How does interest formula change for more than 1 year? _____
        4. Why is compound interest different from simple? _____
        _____

Ch 13 Ls 2 Compound Interest Formula 239

_____ #1 #2 ____ / 5 #3 ____ / 2 R ___ / 8 T ____ / 15 _____
        Name   Checker

#1 1. What is the compound interest formula? _____
    2. What 2 things does compound interest change? _____

_____

#2 1. What's the compound   Mr K invested Rs 4000 at 8% compounded
       interest formula?   4 times a year for 2 years.

           Solve the 1st step.   t = _____ (1 + ———) ‾

           Finish it. Estimate if needed.   _____

_____

                                   Mr W owes Rs 25,000 at 12% compounded
    2. What's the equation?   4 times a year for 2 years.

           Solve the 1st step.   t = _____ (1 + ———) ‾

           Finish it. Estimate if needed.   _____

_____

                                   Mr T owes Rs 15,000 at 12% compounded
    3. What's the equation?   12 times a year for 3 years.

           Solve the 1st step.   t = _____ (1 + ———) ‾

           Finish it. Estimate if needed.   _____

                                                _____

#3 Solve these problems with compound interest.   Calculator? yes no

1. Mr B invested Rs 8000 at 12% compounded 4 times a year for 2 years.

   t = _____(1 + ———) ―

2. Mr D owes Rs 30,000 at 12% compounded 8 times a year for 3 years.

   t = _____(1 + ———) ―

Review 1. What is the compound interest formula? _____   Calculator?
2. What 2 things does compound interest change? _____   yes no

What is happening?

3. $6000(1 + 1.08/4)^8 = t$ Rs 6000 at ___% compounded ___, ___ yr

4. $10{,}000(1 + 1.15/8)^{24} = t$ Rs 10,000 at ___% compounded ___, ___ yr

5. $8000(1 + 0.20/12)^{24} = t$ Rs 8000 at ___% compounded ___, ___ yr

6. $4200(1 + 0.25/8)^{16} = t$ Rs 4200 at ___% compounded ___, ___ yr

7. $1500(1 + 0.30/12)^{36} = t$ Rs 1500 at ___% compounded ___, ___ yr

8. $2000(1 + 0.12/8)^{16} = t$ Rs 2000 at ___% compounded ___, ___ yr

Review Problems   241

_____   #1 #2 ____ / 8   #3 ____ / 5   Total ____ / 13
                Name

#1   1. Mr K deposited Rs 4000 into an account paying   _____   Calculator?
     6% annual interest compounded quarterly. How   yes   no
     much money will be in the account after 2 years? _____
Solve with   __x__
compound   t = _____ (1 + ———)   _____
interest.
                                                                                   _____

   2. If you get a loan for Rs 6000 at 12% annual   _____
      interest compounded monthly for a year, how
      much money will you pay monthly?   _____

                                          __x__
         t = _____ (1 + ———)   _____

                                                              _____

   3. Eva deposited Rs 6500 into an account paying   _____
      2% annual interest compounded monthly. How
      much money will be in the account after 3 yrs?   _____

                                          __x__
         t = _____ (1 + ———)   _____

                                                              _____

   4. Mosholu Bank pays 4% interest that is   _____
      compounded annually. If the original investment
      is Rs 2,000, determine the balance after 1 year.   _____

                                          __x__
         t = _____ (1 + ———)   _____

                                                              _____

   5. Ojas got a loan for Rs 20,000 for his home. He   _____
      pays 3% annual interest compounded monthly.
      How much does he pay monthly?   _____

                                          __x__
         t = _____ (1 + ———)   _____

                                                              _____

Story Problems.

#2

1. If you take a car loan for Rs 30,000 at 7% compounded quarterly. What will your balance be at the end of those five years?

Calculator? yes no

$$t = \underline{\qquad}(1 + \underline{\qquad})^{\underline{\quad}\times\underline{\quad}}$$

2. If you take a car loan for Rs 10,000 at 5% compounded monthly. What will your balance be at the end of those four years?

$$t = \underline{\qquad}(1 + \underline{\qquad})^{\underline{\quad}\times\underline{\quad}}$$

3. If Rs 4000 is invested in an account paying 3% interest compounded monthly, what is the balance after 4 years?

$$t = \underline{\qquad}(1 + \underline{\qquad})^{\underline{\quad}\times\underline{\quad}}$$

4. If Rs 6000 is invested in an account paying 4% interest compounded quarterly, what is the balance after 3 years?

$$t = \underline{\qquad}(1 + \underline{\qquad})^{\underline{\quad}\times\underline{\quad}}$$

5. Tom takes a house loan for Rs 5,000 at 5% compounded quarterly. What will your balance be at the end of those two years?

$$t = \underline{\qquad}(1 + \underline{\qquad})^{\underline{\quad}\times\underline{\quad}}$$

Ch 13 Ls 3 Premade Equations. 243

_____ #1 #2 ____/ 6   #3 ____/ 3   R ____/ 6 T ____/ 15 _____
           Name   Checker

#1 1. How are manufacturing problems different from other problems? _____

_____

2. What formula does an equation use to find the highest point of a graph? _____
3. How does a problem find where an equation hits 0 as a point? _____

#2 1. What's the compound   Rs 5000 compounded 4 times a year
        interest formula?   at 16% interest for 2 year.

            Solve the 1st step. $t = \_\_\_\_(1 + \_\_\_\_) - ^{x} \_\_$

        Finish it. Estimate if needed.

        2. What's the compound   Rs 12,000 compounded 8 times a year
        interest formula?   at 12% interest for 3 year.

            Solve the 1st step. $t = \_\_\_\_(1 + \_\_\_\_) - ^{x} \_\_$

        Finish it. Estimate if needed.

        3. What's the compound   Rs 2000 compounded 12 times a year
        interest formula?   at 18% interest for 4 year.

            Solve the 1st step. $t = \_\_\_\_(1 + \_\_\_\_) - ^{x} \_\_$

        Finish it. Estimate if needed.

#3 1. A trapezoid has area of A = h(b1 + b2) / 2. What is the area of trapezoid with height of 6 inches, b1 is 6 inches, and b2 is 12 in?

A = 6 (6 + 12) / 2

Calculator? yes   no

Solve these Premade Equations.

2. The area of a hexagon is A = 1/2 a P where the closest point is 6 and perimeter is 24 m.

$A = \frac{1}{2} \sqrt{6} (24)$

3. A sphere's surface area is 4pi r^2 or 12.6 r^2. A sphere has a radius 5 inches. How much paper does it take to cover it?

$12.6 (5)^2 = s$

4. A triangular sign has a base is 3 feet less than twice the height. The law states the sign cannot be more than 21 sq m. The equation is 1/2 (2h - 3) h = 21. How tall is the sign?

$\frac{1}{2} (2h - 3) h = 21$

5. Suppose 320 tickets were sold for a game for a total of Rs 12,400. If adult tickets cost Rs 50 each and children's tickets cost Rs 30, how many tickets were sold?

Review 1. How are manufacturing problems different from other problems? _____

Calculator? yes   no

2. What formula does an equation use to find the highest point of a graph? _____

3. How does a problem find where an equation hits 0 as a point? _____

Review Problems 245

_____ #1 to #3 ____ / 11 #2 ____ / 5 Total ____ / 16
           Name

1. All the Money Formula _____
2. Compound Interest _____
3. Premade Equations _____

#2 Decide what's happening, then solve them.   Calculator?
                                                              yes   no

1.   8000 × 1.05 = t   7000 × 1.10 = t
    ___ at ___ interest for 1 yr   ___ at ___ interest for 1 yr
    _____   _____

2. $9000(1 + 1.20)^2 = t$ $3500(1 + 1.50)^2 = t$
    ___ at ___ interest for ___ years   ___ at ___ interest for ___ years
    _____   _____

3. $2000(1 + 1.70)^2 = t$ $3500(1 + 2.00)^2 = t$
    ___ at ___ interest for ___ years   ___ at ___ interest for ___ years
    _____   _____

                                                              Calculator?
#3 1. Mr B invested Rs 5200 at 12%   yes   no
      compounded 4 times a year for 3 years.   _____

      t = _____ (1 + ———) $^{\_x\_}$
                                               _____

                                               _____

2. Mr D owes Rs 8600 at 12% compounded
   8 times a year for 2 years.   _____

      t = _____ (1 + ———) $^{\_x\_}$
                                               _____

                                               _____

                                               _____

Story problems.

1. Ojas owes Rs 20,000 at 16% compounded 4 times a year for 2 years.

   t = _____ (1 + ___) — $\frac{x}{}$ —

   Calculator? yes   no

2. A sphere has a radius of 6 cm. It has a volume 4/3 pi r^3 or 4.1 r cubed. How much water will it take to fill the sphere up?

   6 cm

   4.1 (6^3) = v

3. Zara owes Rs 9000 at 16% compounded 8 times a year for 8 years.

   t = _____ (1 + ___) — $\frac{x}{}$ —

4. What is the velocity of a car that traveled a total of 120 kilometers north in 1.5 hours?

   r x 1.5 hr = 120 km

5. A washer-dryer combination costs Rs 2720. If the washer costs Rs 800 more than the dryer, what does each appliance cost?

   x + (x + 80) = 2720

6. Miul has a board that is 120 centimeters long. He wishes to cut the board into two pieces so that one piece will be 8 in. longer than the other. How long are the pieces?

   x + (x + 8) = 120

Ch 14 Ls 1 Pythagorean Theorem   247

_____ #1 #2 ____ / 7 #3 ____ / 9 R ____ / 6 T ____ / 22 _____
          Name   Checker

#1 1. What is the Pythagorean Theorem? _____
      2. What does Pythagorean Theorem do? _____
      3. What kind of triangles does Pythagorean work with? _____
      4. How can you tell if it makes a right triangle? _____

#2 1. How long is A? What's the equation?   6 / 4 / a

   ___^{2} + ___^{2} = ___^{2} Solve the next step.

   ___ + ___ = ___   Estimate to 10ths.

   _____

   _____

   2. How long is B? What's the equation?   7 / 2 / b

   ___^{2} + ___^{2} = ___^{2} Solve the next step.

   ___ + ___ = ___   Estimate to 10ths.

   _____

   _____

   3. Is this a right triangle? **A is 4   B is 6   C is 7**

   ___ + ___ = ___   What's the 1st step?

   _____   Right triangle or not?

   _____

#3 Solve with Pythagorean Thereom.   Calculator? yes   no

1.
[triangle: c, legs 6 and 3]   [triangle: c, legs 7 and 3]   [triangle: c, legs 8 and 3]

___ + ___ = ___   ___ + ___ = ___   ___ + ___ = ___

_____   _____   _____

_____   _____   _____

Triangles are not in drawn in shape.

2.
[triangle: a, sides 5 and 4]   [triangle: a, sides 7 and 5]   [triangle: a, sides 9 and 7]

___ + ___ = ___   ___ + ___ = ___   ___ + ___ = ___

_____   _____   _____

_____   _____   _____

3.
[triangle: b, sides 6 and 3]   [triangle: b, sides 5 and 4]   [triangle: b, sides 8 and 3]

___ + ___ = ___   ___ + ___ = ___   ___ + ___ = ___

_____   _____   _____

_____   _____   _____

Review 1. What is the Pythagorean Theorem? _____   Calculator? yes   no

2. What does Pythagorean Theorem do? _____

3. What kind of triangles does Pythagorean work with? _____

4. How can you tell if it makes a right triangle? _____

5. A is 4 B is 6 C is 7.2   A is 5 B is 7 C is 9

Is this a right triangle or not?   ___ + ___ = ___   ___ + ___ = ___

_____ YES NO   _____ YES NO

Ch 14 Ls 2 Pythagorean Theorem Story Problems   249

_____ #1 #2 ____ / 7 #3 #4 ____ / 6 R ___ / 4 T ____ / 17 _____
         Name   Checker

#1 1. What are the 3 parts of the triangle in the cat in the tree problem? _____

_____

2. How does Pythagorean Theroem use rate formula? _____

3. Name 2 things a problem needs to use rate with Pythagorean formula. _____

_____

#2 1. What are the   Mr K drove east at 30 kph for 0.5 hrs. Tom drove
         rate formulas?   south 70 kph for 0.5 hrs. How far are they apart?

                Mr K   Tom   What is the
                ____ x ____ = ____   ____ x ____ = ____   equation?

                                              _____   What's the 1st step?

                                              _____   What's the next step?

                                              _____   What's the last step?

                                              _____   Find the answers.

                                              _____

         2. What are the   Two cars with CB radios have a range of 30 kilometers.
            rate formulas?   Radio #1 heads east at 30 kph and #2, south at
                               40 kph. How long will they be able to talk?

                #1 car   #2 car   What is the
                ____ x ____ = ____   ____ x ____ = ____   equation?

                                              _____   What's the 1st step?

                                              _____   What's the next step?

                                              _____   What's the last step?

                                              _____   Find the answers.

                                              _____

#3 1. A 20 m ladder leans against a wall. How high does it reach if the foot is 3 m from the wall?

Equation ___ + ___ = ___

Calculator? yes   no

Draw the problem and solve it.

2. A baseball park has a length of 30 m from home plate to first base. How far is it from home plate to second base?

Equation ___ + ___ = ___

Decide if is right triangle.

1. A is 4 B is 6 C is 7.7

___ + ___ = ___

A is 5 B is 8 C is 9.4

___ + ___ = ___

2. A is 2 B is 3 C is 3.6

___ + ___ = ___

A is 3 B is 5 C is 5.4

___ + ___ = ___

Review 1. What are the 3 parts of the triangle in the cat in the tree problem? _____

Calculator? yes   no

2. Where is 11 ft in the problem? _____
3. How does Pythagorean Theroem use rate formula? _____
4. Name 2 things a problem needs to use rate with Pythagorean formula. _____

Ch 14 Ls 3 Rate Formulas and Pythagorean Theorem   251

_____   #1 #2 ____/ 6   #3 ____/ 4   R ____/ 4   T ____/ 14 _____
                Name   Checker

#1 1. What is the formula for a circle? _____

2. Why can Pythagorean make the formula for a circle? _____
_____

3. What are axis points? _____

4. What is a ballpark estimate? _____

#2 1. What are the rate formulas? $a^2 + 2^2 = 3^2$ What's the 1st step?

                                    _____   What's the next step?

                                    _____   Graph it.

1. What are the rate formulas? $3^2 + b^2 = 4^2$ What's the 1st step?

                                    _____   What's the next step?

                                    _____   Graph it.

#3 1. $a^2 + 5^2 = 6^2$ $3^2 + b^2 = 6^2$ Calculator? yes no

Solve the 1st step. _____   _____

Estimate each radius. _____   _____

Graph it _____   _____

2. $4^2 + 5^2 = c^2$ $7^2 + b^2 = 9^2$

Solve the 1st step. _____   _____

Estimate each radius. _____   _____

Graph it _____   _____

Review 1. What is the formula for a circle? _____

2. Why can Pythagorean make the formula for a circle? _____

3. What are axis points? _____

4. What is a ballpark estimate? _____

Ch 14 Ls 4 Simultaneous Equations with Quadratic Shapes 253

_____ #1 #2 ____ / 8 #3 ____ / 4 R ____ / 4 T ____ / 16 _____
       Name   Checker

#1 1. What do you need to substitute equations? _____
       2. How does the 1st equation substitute into the other? _____
       3. After you find 1 variable, what happens? _____
       4. Substitute it. What's the new equation? $x^2 - 1 = y$
                                                     $-x + y = 1$

                           What's the 1st step?   _____

                           What's the answer?   _____

                           Use equation #2.
                                                   _____

                           Solve it. What's the point? _____

                                                   _____

#2 1. How do you eliminate a variable? _____
       2. What if no terms can be subtracted? _____
       3. How do you subtract to eliminate a variable? _____
       4. Eliminate it. What's the new equation? $x^2 - y = 9$
                                                     $-x + y = 1$

                           What's the 1st step?   _____

                           What's the answer?   _____

                           Use equation #2.
                                                   _____

                           Solve it. What's the point? _____

                                                   _____

#3 1. Substitute it. What's the new equation?

$3x^2 - 2 = y$
$2x + y = 1$

$6x^2 - 1 = y$
$3x - y = 4$

Calculator?   yes   no

What's the 1st step? _____   _____

   What's the answer? _____   _____

     Use equation #2. _____   _____

       Solve it.
      What's the point? _____   _____

_____   _____

2. Eliminate it. What's the new equation?

$y = 2x^2 + 3x - 8$
$y = 2x^2 - 4x + 5$

$x^2 + y^2 = 7$
$4x^2 - y^2 = 8$

What's the 1st step? _____   _____

   What's the answer? _____   _____

     Use equation #2. _____   _____

       Solve it.
      What's the point? _____   _____

_____   _____

Review 1. What do you need to substitute equations? _____

   2. How does the 1st equation substitute into the other? _____

   3. After you find 1 variable, what happens? _____

   4. How do you eliminate a variable? _____

   5. What if no terms can be subtracted? _____

   6. How do you subtract to eliminate a variable? _____

Review Problems   255

_____ #1 #2 #3 _____ / 14 #4 ___/ 6 Total ____/ 20
         Name

1. Pythagorean Formula _____
2. Rate Problems _____
3. Circuference of a Circle _____
4. Area of a Circle _____
5. Substitute Equations _____
6. Eliminate Equations _____

#2 Solve with Pythagorean Thereom.   Calculator?
                                                       yes   no

1. c ◢ 4   8 ◢ b   9 ◢ 2
    ‾7‾   ‾6‾   ‾a‾
   __ + __ = __   __ + __ = __   __ + __ = __

   _____   _____   _____
   _____   _____   _____

2. c ◢ 5   7 ◢ b   12 ◢ 3
    ‾7‾   ‾6‾   ‾a‾

Triangles are not
in drawn in shape. __ + __ = __   __ + __ = __   __ + __ = __

   _____   _____   _____
   _____   _____   _____

                                                       Calculator?
#3 1. Substitute it. What's $3x^2 - 2 = y$ $6x^2 - 1 = y$ yes   no
       the new equation? $x + y = 1$ $x - y = 1$

    What's the 1st step?   _____   _____

    What's the answer?   _____   _____

    Use equation #2.   _____   _____
       Solve it.
    What's the point?   _____   _____

                              _____   _____
                              _____   _____

#4 1. A ship travels west 30 km to Ada Island, then north 12 km to B Island. How far is it from the starting point?

___ + ___ = ___

Calculator?
yes   no

Make an equation before solving it.

2. A 30 m ladder leans against a wall. How high does it reach if the foot is 5 m from the wall?

___ + ___ = ___

3. A baseball park has a length of 25 m from home plate to first base. How far is it from home plate to second base?

___ + ___ = ___

4. A toy fishing rod is 1 meter long. It has a 0.6 meter line with a rubber duck on the end. With the duck level with the base, how far out does it extend?

___ + ___ = ___

5. You put up a banner for a party that is 8 meters across and 4 meters up the right side. How far is the hyotenuse for the triangle?

___ + ___ = ___

6. You leave your house and run 5 km west. Then, you turn and run 2 km sourth. How far are you from home?

___ + ___ = ___

Ch 15 Ls 1 Make it Change it, Tank Problems.   257

_____ #1 #2 ____/ 8 #3 ____/ 3 R ___/ 5 T ____/ 16 _____
          Name   Checker

#1 1. How does it start for the tank problem? _____
    2. What changed the equation? _____
    3. What are 2 ways to solve it? _____
    4. What's the   A company makes a 10 centimeter pipe who's radius is 5 cm
       formula?   wide. How does an cm in radius change the volume?

                              _____   What is the
                                                            equation?

                              _____   Make a variable.

                              _____   1st step?

                              _____   Find the answer.

                              _____

#2 1. What formula finds the velocity of water in a pipe? _____
    2. What does Q and A stand for? _____
    3. What does v stand for? _____
    4. What's the   What's the flow of water in an area pipe of 0.18 meters and
       formula?   velocity of 0.20 m/s?

                              _____   What is the
                                                            equation?

                              _____   Multiply it.

                              _____   How many liters it it?

                              _____

#3 1. The Pipe Company makes several pipes for companies. How many liters flow in a 12 Lpm in a 10 centimeter pipe?

Equation _____

1st step? _____

Next step? _____

Last step? _____

Answer _____

Calculator? yes   no

2. Great City bought a 15 centimeter pipe. How many liters flow in a 12 Lpm with a 15 centimeter pipe?

Equation _____

1st step? _____

Next step? _____

Last step? _____

Answer _____

3. The state bought a 20 centimeter pipe. How many liters flow in a 8 Lpm in a pipe?

Equation _____

1st step? _____

Next step? _____

Last step? _____

Answer _____

Review

1. How does it start for the tank problem? _____

2. What changed the equation? _____

3. What are 2 ways to solve it? _____

4. What formula finds the velocity of water in a pipe? _____

5. What does Q and A stand for? _____

6. What does v stand for? _____

Calculator? yes   no

Ch 15 Ls 2 Box and Mat Problems 259

_____ #1 #2 ____/ 8 #3 ____/ 2 R ___/ 6 T ___/ 16 _____
           Name   Checker

#1 1. What equation started the garden problem? _____
       2. How did making the garden bigger change the equation? _____
       3. Name 2 ways the problem solved. _____

       4. What's the   A garden is 4 m long and 2 m wide. Keep the same shape,
          formula?   but add an edge. If the edge is 2 m wide, What's the area?

                                                            _____ What is the
                                                                              equation?

                                                            _____ Make a variable.

                                                            _____ 1st step?

                                                            _____ Find the answer.

                                    _____

#2 1. What is the 1st step to the box problem? _____
       2. How did the 2 inch corners change X? _____ | 4 cm 4 cm |
       3. What is length, width, and height here? _____

       4. What's the   A square pan will will have 4 centimeter sides. What size
          formula?   sheet will make a pan with volume of 1600 cubic cm? | 4 cm 4 cm |

                                                            _____ What is the
                                                                              equation?

                                                            _____ Make a variable.

                                                            _____ 1st step?

                                                            _____ Find the answer.

                                    _____

#3 Answer each with quadrilateral instructions.   Calculator?
                                                      yes   no

1. **A pool is 3 m long and 2 m wide. Keep the same shape, but add an edge. If the edge is 1 m wide, What's the area?**

   Equation _____

   1st step? _____

   Next step? _____

   Last step? _____

   Answer. _____

   Other answer? _____

2. **A driveway is 2 m wide and 30 m long. Keep the same shape, but add an edge. If the edge is 1 m wide, What's the area?**

   Equation _____

   1st step? _____

   Next step? _____

   Last step? _____

   Answer. _____

   Other answer? _____

   _____

Review 1. What equation started the garden problem? _____

2. How did making the garden bigger change the equation? _____

3. Name 2 ways the problem solved. _____

4. What is the 1st step to the box problem? _____   [4 cm 4 cm]

5. How did the 2 inch corners change X? _____   [4 cm 4 cm]

6. What is length, width, and height here? _____

Review Problems 261

_____   #1 #2 ____ / 8 #3 ____ / 5 Total ____ / 13
Name

#1 1. Tank Problem _____
    2. Garden Problem _____
    3. Pan Problem _____

#2 1. How many liters flow in a 10 Lpm through a   Calculator?
       8 cm pipe?   yes   no

Solve these
story problems.

    2. How many liters flow in a 15 Lpm through a
       10 centimeter pipe?

    3. How many liters flow in a 30 Lpm through a
       1 meter pipe?

    4. How many liters flow in a 50 Lpm through a
       1.5 meter pipe?

    5. How many liters flow in a 100 Lpm through a
       2 meter pipe?

#3

1. The height of a triangle is 5 units less than the length of the base. If the area of the triangle is 50 square units, find the length of its base and height.

Calculator?
yes   no

2. A uniform border is to be placed around an 6 cm by 8 cm picture. If the total area including the border must be 99 square cm, then how wide should the border be?

3. A garden is 12 m long and 14 m wide. Keep the same shape, but add an edge. If the edge is 5 m wide, what's the area?

4. A box can be made by cutting out the corners and folding up the edges of a square sheet of cardboard. Each side has a height of 3 centimeters. What is the length of each side of the cardboard sheet if the volume of the box is to be 432 cubic cms?

5. A square pan will will have 6 cm sides. What size sheet will make a pan with volume of 726 cubic centimeters?

Ch 15 Ls 3 Ohm's Law/ Resistance Formula   263

_____ #1 #2 ____/ 10 #3 ____/ 4 R ____/ 10 T ____/ 22 _____
              Name   Checker

#1   1. What is Ohm's Law? _____

   2. What is current measured in? _____

   3. What is resistance measured in? _____

   4. What is voltage measured in? _____

   5. Make an equation.   **What is the voltage for 0.2 amps at 110 Ohms.**

              What is the answer? _____

   6. Make an equation.   **What is the voltage for 3 amps at 18 Ohms.**

              What is the answer? _____

#2   1. What is the resistance formula? _____

   2. What is the resistance formula with a square root? _____

   3. What is the square of 50? $P = \dfrac{50^2}{100}$ ohms

              What's the answer? _____

   4. What is the square of 20? $P = \dfrac{20^2}{60}$ ohms

              What's the answer? _____

#3

1. How much current flows through a light bulb that has 5 ohms of resistance when 110 volts are applied to it? _____

   Calculator? yes   no

2. How much current flows through a heating element that has a resistance of 40 ohms when 220 volts are applied to it? _____

3. How much current flows through an electric toaster that has a resistance of 30 ohms when 120 volts are applied to it? _____

4. What is the resistance of a 60 Watt light bulb that allows 2 Amps to flow when 120 V is applied to it? _____

5. What is the resistance of a heating element that allows 15 Amps to flow when 220 V is applied to it? _____

6. What battery voltage would cause 3 Amps to flow through a 300 ohm resistor? _____

7. How much voltage is required to make 15 Amps flow through a 7.5 ohm resistance? _____

Review

1. What is the formula for electrical power? _____

   Calculator? yes   no

2. What is current measured in? _____

3. 3. What is voltage measured in? _____

4. What is power measured in? _____

5. What is the resistance formula? _____

6. What is the resistance formula with a square root? _____

Ch 15 Ls 4 Science: Pendulum/Parachute   265

_____ #1 #2 ____/9 #3 ____/8 R ____/6 T ____/23 _____
Name   Checker

#1 1. What formula finds time a pendulum makes a full pass? _____
   2. What does L stand for? _____
   3. What's the number for pi squared? _____
   4. Make an equation. **A pendulum is 5 m long. Find time for 1 pass.**

   $$t = \sqrt{_____}$$ What is the answer?

   $$t = \sqrt{_____} = $$

   5. Make an equation. **A pendulum is 10 m long. Find time for 1 pass.**

   $$t = \sqrt{_____}$$ What is the answer?

   $$t = \sqrt{_____} = $$

#2 1. What is the parachute formula? _____
   2. What does M stand for? _____
   3. What does L stand for? _____

   4. Make an equation. **You and a parachute weigh 60 kg. Find the parachute size
      If you intend a 3 meter/sec landing.**

   $$D = \sqrt{_____}$$

   $$D = \sqrt{_____} = $$

                                         What is the answer?

#3

1. $t = \sqrt{\dfrac{9.9(5) \text{ m}}{8}}$ $t = \sqrt{\dfrac{9.9(8) \text{ m}}{8}}$ Calculator? yes   no

Use the pendulum and parachute formula.

_____   _____

_____   _____

2. $t = \sqrt{\dfrac{9.9(12) \text{ m}}{8}}$ $t = \sqrt{\dfrac{9.9(14) \text{ m}}{8}}$

_____   _____

_____   _____

3. $D = \sqrt{\dfrac{74(100) \text{ m}}{5.5(2^2)}}$ $D = \sqrt{\dfrac{74(80) \text{ m}}{5.5(4^2)}}$

$D = \sqrt{\dfrac{}{}}$ $D = \sqrt{\dfrac{}{}}$

$D = \sqrt{}$ $D = \sqrt{}$

D = _____ m/sec   D = _____ m/sec

Review
1. What formula finds time a pendulum makes a full pass? _____
2. What does L stand for? _____
3. What's the number for pi squared? _____
4. What is the parachute formula? _____
5. What does M stand for? _____
6. What does L stand for? _____

Review Problems   267

_____ #1 #2 ____ / 12   #3 #5 ____ / 8   Total ____ / 20
           Name

#1 1. Power Formula _____

    2. Resistance Formula _____

    3. Pendulum Problem _____

    4. Parachute Probem _____

                                                                    Calculator?
#2 1. You have a 3600 watt element, 240 volt water   _____   yes   no
       heater circuit, how many amps?
Make an equation   _____
before solving.

    2. A 100 W light bulb draws 833 mA when the   _____
       rated voltage is applied. What is the rated
       voltage?   _____

    3. A juicer uses 90 W of Power and the current is   _____
       4.5 A. How many volts are necessary to run it?
                                                         _____

    4. Find the current through a 12-ohm resistive   _____
       circuit when 24 volts is applied
                                                         _____

    5. Find the resistance of a circuit that draws 0.06 _____
       amperes with 12 volts applied
                                                         _____

    6. A small appliance is rated at a current of 10
       amps and a voltage of 120 volts, the power   _____
       rating would be _____ Watts. (P = I V)
                                                         _____

    7. If a blender is plugged into a 110 V outlet that
       supplies 2.7 A of current, what amount of   _____
       power is used by the blender?
                                                         _____

    8. If a clock uses 2 W of power from a 1.5 V
       battery, what amount of current is supplying   _____
       the clock?
                                                         _____

#4 Use the pendulum formula for these.   Calculator? yes no

1. $t = \sqrt{\dfrac{9.9(2)}{8}}$ m $t = \sqrt{\dfrac{9.9(6)}{8}}$ m

2. $t = \sqrt{\dfrac{9.9(7)}{8}}$ m $t = \sqrt{\dfrac{9.9(10)}{8}}$ m

#5 Use the parachute formula.   Calculator? yes no

1. $v = \sqrt{\dfrac{2(100)}{1(1)60}}$ $v = \sqrt{\dfrac{2(90)}{1(1)40}}$

2. $v = \sqrt{\dfrac{2(120)}{1(1)70}}$ $v = \sqrt{\dfrac{2(80)}{1(1)50}}$

www.ingramcontent.com/pod-product-compliance
Lightning Source LLC
Chambersburg PA
CBHW020734180526
45163CB00001B/226